Chapman & Hall/CRC
Computer Science and Data Analysis Series

The interface between the computer and statistical sciences is increasing, as each discipline seeks to harness the power and resources of the other. This series aims to foster the integration between the computer sciences and statistical, numerical, and probabilistic methods by publishing a broad range of reference works, textbooks, and handbooks.

SERIES EDITORS
David Blei, Princeton University
David Madigan, Rutgers University
Marina Meila, University of Washington
Fionn Murtagh, Royal Holloway, University of London

Proposals for the series should be sent directly to one of the series editors above, or submitted to:

Chapman & Hall/CRC
Taylor and Francis Group
3 Park Square, Milton Park
Abingdon, OX14 4RN, UK

Published Titles

Semisupervised Learning for Computational Linguistics
Steven Abney

Visualization and Verbalization of Data
Jörg Blasius and Michael Greenacre

Design and Modeling for Computer Experiments
Kai-Tai Fang, Runze Li, and Agus Sudjianto

Microarray Image Analysis: An Algorithmic Approach
Karl Fraser, Zidong Wang, and Xiaohui Liu

R Programming for Bioinformatics
Robert Gentleman

Exploratory Multivariate Analysis by Example Using R
François Husson, Sébastien Lê, and Jérôme Pagès

Bayesian Artificial Intelligence, Second Edition
Kevin B. Korb and Ann E. Nicholson

Chapman & Hall/CRC
Computer Science and Data Analysis Series

Exploratory Multivariate Analysis by Example Using R

François Husson
Sébastien Lê
Jérôme Pagès

CRC Press
Taylor & Francis Group
Boca Raton London New York

CRC Press is an imprint of the
Taylor & Francis Group, an **informa** business
A CHAPMAN & HALL BOOK

CRC Press
Taylor & Francis Group
6000 Broken Sound Parkway NW, Suite 300
Boca Raton, FL 33487-2742

First issued in paperback 2020

© 2017 by Taylor & Francis Group, LLC
CRC Press is an imprint of Taylor & Francis Group, an Informa business

No claim to original U.S. Government works

ISBN 13: 978-0-367-65802-1 (pbk)
ISBN 13: 978-1-1381-9634-6 (hbk)

**Visit the Taylor & Francis Web site at
http://www.taylorandfrancis.com**

**and the CRC Press Web site at
http://www.crcpress.com**

Contents

Preface

Qu'est-ce que l'analyse des données ? (English: What is data analysis?)

As it is usually understood in France, and within the context of this book, the expression *analyse des données* reflects a set of statistical methods whose main features are to be multidimensional and descriptive.

The term *multidimensional* itself covers two aspects. First, it implies that observations (or, in other words, individuals) are described by several variables. In this introduction we restrict ourselves to the most common data, those in which a group of individuals is described by one set of variables. But, beyond the fact that we have many values from many variables for each observation, it is the desire to study them simultaneously that is characteristic of a multidimensional approach. Thus, we will use those methods each time the notion of profile is relevant when considering an individual, for example, the response profile of consumers, the biometric profile of plants, the financial profile of businesses, and so forth.

From another point of view, the interest of considering values of individuals for a set of variables in a global manner lies in the fact that these variables are linked. Let us note that studying links between all the variables taken two-by-two does not constitute a multidimensional approach in the strict sense. This approach involves the simultaneous consideration of all the links between variables taken two-by-two. That is what is done, for example, when highlighting a synthetic variable: such a variable represents several others, which implies that it is linked to each of them, which is only possible if they are themselves linked two-by-two. The concept of synthetic variable is intrinsically multidimensional and is a powerful tool for the description of an individuals × variables table. In both respects, it is a key concept within the context of this book.

One last comment about the term *analyse des données* since it can have at least two meanings — the one defined previously and another broader one that could be translated as "statistical investigation." This second meaning is from a user's standpoint; it is defined by an objective (to analyse data) and says nothing about the statistical methods to be used. This is what the English term *data analysis* covers. The term *data analysis*, in the sense of a set of descriptive multidimensional methods, is more of a French statistical point of view. It was introduced in France in the 1960s by Jean-Paul Benzécri and the adoption of this term is probably related to the fact that these multivariate methods are at the heart of many "data analyses."

To Whom Is This Book Addressed?

This book has been designed for scientists whose aim is not to become statisticians but who feel the need to analyse data themselves. It is therefore addressed to practitioners who are confronted with the analysis of data. From this perspective it is application oriented; formalism and mathematics writing have been reduced as much as possible while examples and intuition have been emphasised. Specifically, an undergraduate level is quite sufficient to capture all the concepts introduced.

On the software side, an introduction to the R language is sufficient, at least at first. This software is free and available on the Internet at the following address: `http://www.r-project.org/`.

Content and Spirit of the Book

This book focuses on four essential and basic methods of multivariate exploratory data analysis, those with the largest potential in terms of applications: principal component analysis (PCA) when variables are quantitative, correspondence analysis (CA) and multiple correspondence analysis (MCA) when variables are categorical and hierarchical cluster analysis. The geometric point of view used to present all these methods constitutes a unique framework in the sense that it provides a unified vision when exploring multivariate data tables. Within this framework, we will present the principles, the indicators, and the ways of representing and visualising objects (rows and columns of a data table) that are common to all those exploratory methods. From this standpoint, adding supplementary information by simply projecting vectors is commonplace. Thus, we will show how it is possible to use categorical variables within a PCA context where variables that are to be analysed are quantitative, to handle more than two categorical variables within a CA context where originally there are two variables, and to add quantitative variables within an MCA context where variables are categorical. More than the theoretical aspects and the specific indicators induced by our geometrical viewpoint, we will illustrate the methods and the way they can be exploited using examples from various fields, hence the name of the book.

Throughout the text, each result correlates with its R command. All these commands are accessible from FactoMineR, an R package developed by the authors. The reader will be able to conduct all the analyses of the book as all the datasets (as well as all the lines of code) are available at the following website address: `http://factominer.free.fr/bookV2`. We hope that with this book, the reader will be fully equipped (theory, examples, software) to confront multivariate real-life data.

Note on the Second Edition

There were two main reasons behind the second edition of this work. The first was that we wanted to add a chapter on viewing and improving the graphs produced by the FactoMineR software. The second was to add a section to each chapter on managing missing data, which will enable users to conduct analyses from incomplete tables more easily.

The authors would like to thank Rebecca Clayton for her help in the translation.

1

Principal Component Analysis (PCA)

1.1 Data — Notation — Examples

Principal component analysis (PCA) applies to data tables where rows are considered as individuals and columns as quantitative variables. Let x_{ik} be the value taken by individual i for variable k, where i varies from 1 to I and k from 1 to K.

Let $\bar{x}_k$ denote the mean of variable k calculated over all individual instances of I:

$$\bar{x}_k = \frac{1}{I} \sum_{i=1}^{I} x_{ik} \; ,$$

and s_k the standard deviation of the sample of variable k (uncorrected):

$$s_k = \sqrt{\frac{1}{I} \sum_{i=1}^{I} (x_{ik} - \bar{x}_k)^2} \; .$$

Data subjected to a PCA can be very diverse in nature; some examples are listed in Table 1.1.

This first chapter will be illustrated using the "orange juice" dataset chosen for its simplicity since it comprises only six statistical individuals or observations. The six orange juices were evaluated by a panel of experts according to seven sensory variables (odour intensity, odour typicality, pulp content, intensity of taste, acidity, bitterness, sweetness). The panel's evaluations are summarised in Table 1.2.

1.2 Objectives

The data table can be considered either as a set of rows (individuals) or as a set of columns (variables), thus raising a number of questions relating to these different types of objects.

TABLE 1.1

Some Examples of Datasets

Field	Individuals	Variables	x_{ik}
Ecology	Rivers	Concentration of pollutants	Concentration of pollutant k in river i
Economics	Years	Economic indicators	Indicator value k for year i
Genetics	Patients	Genes	Expression of gene k for patient i
Marketing	Brands	Measures of satisfaction	Value of measure k for brand i
Pedology	Soils	Granulometric composition	Content of component k in soil i
Biology	Animals	Measurements	Measure k for animal i
Sociology	Social classes	Time by activity	Time spent on activity k by individuals from social class i

TABLE 1.2

The Orange Juice Data

	Odour intensity	Odour typicality	Pulp	Intensity of taste	Acidity	Bitter-ness	Sweet-ness
Pampryl amb.	2.82	2.53	1.66	3.46	3.15	2.97	2.60
Tropicana amb.	2.76	2.82	1.91	3.23	2.55	2.08	3.32
Fruvita fr.	2.83	2.88	4.00	3.45	2.42	1.76	3.38
Joker amb.	2.76	2.59	1.66	3.37	3.05	2.56	2.80
Tropicana fr.	3.20	3.02	3.69	3.12	2.33	1.97	3.34
Pampryl fr.	3.07	2.73	3.34	3.54	3.31	2.63	2.90

1.2.1 Studying Individuals

Figure 1.1 illustrates the types of questions posed during the study of individuals. This diagram represents three different situations where 40 individuals are described in terms of two variables: j and k. In graph A, we can clearly identify two distinct classes of individuals. Graph B illustrates a dimension of variability which opposes extreme individuals, much like graph A, but which also contains less extreme individuals. The cloud of individuals is therefore long and thin. Graph C depicts a more uniform cloud (i.e., with no specific structure).

Interpreting the data depicted in these examples is relatively straightforward as they are two dimensional. However, when individuals are described by a large number of variables, we require a tool to explore the space in which these individuals evolve. Studying individuals means identifying the similarities between individuals from the point of view of all the variables. In other words, to provide a typology of the individuals: which are the most similar individuals (and the most dissimilar)? Are there groups of individuals which

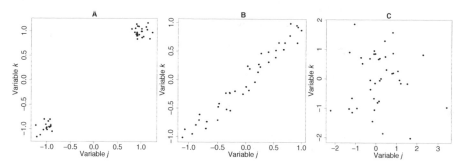

FIGURE 1.1

Representation of 40 individuals described by two variables: j and k.

are homogeneous in terms of their similarities? In addition, we should look for common dimensions of variability which oppose extreme and intermediate individuals.

In the example, two orange juices are considered similar if they were evaluated in the same way according to all the sensory descriptors. In such cases, the two orange juices have the same main dimensions of variability and are thus said to have the same sensory "profile." More generally, we want to know whether or not there are groups of orange juices with similar profiles, that is, sensory dimensions which might oppose extreme juices with more intermediate juices.

1.2.2 Studying Variables

Following the approach taken to study the individuals, might it also be possible to interpret the data from the variables? PCA focuses on the linear relationships between variables. More complex links also exist, such as quadratic relationships, logarithmics, exponential functions, and so forth, but they are not studied in PCA. This may seem restrictive, but in practice many relationships can be considered linear, at least for an initial approximation.

Let us consider the example of the four variables (j, k, l, and m) in Figure 1.2. The clouds of points constructed by working from pairs of variables show that variables j and k (graph A) as well as variables l and m (graph F) are strongly correlated (positively for j and k and negatively for l and m). However, the other graphs do not show any signs of relationships between variables. The study of these variables also suggests that the four variables are split into two groups of two variables, (j, k) and (l, m), and that, within one group, the variables are strongly correlated, whereas between groups, the variables are uncorrelated. In exactly the same way as for constructing groups of individuals, creating groups of variables may be useful with a view to synthesis. As for the individuals, we identify a continuum with groups of both very unusual variables and intermediate variables, which are to some extent linked to both groups. In the example, each group can be represented by one single variable as the variables within each group are very strongly correlated. We refer to these variables as synthetic variables.

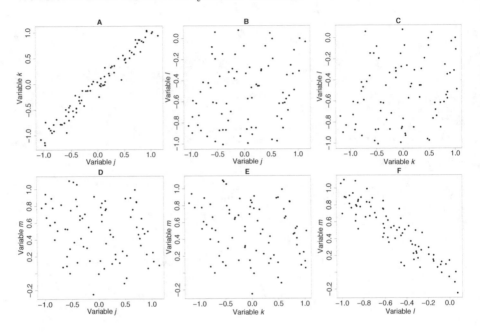

FIGURE 1.2
Representation of the relationships between four variables: j, k, l, and m, taken two-by-two.

When confronted with a very small number of variables, it is possible to draw conclusions from the clouds of points, or from the correlation matrix which groups together all of the linear correlation coefficients $r(j,k)$ between the pairs of variables. However, when working with a great number of variables, the correlation matrix groups together a large quantity of correlation coefficients (190 coefficients for $K = 20$ variables). It is therefore essential to have a tool capable of summarising the main relationships between the variables in a visual manner. The aim of PCA is to draw conclusions from the linear relationships between variables by detecting the principal dimensions of variability. As you will see, these conclusions will be supplemented by the definition of the synthetic variables offered by PCA. It will therefore be easier to describe the data using a few synthetic variables rather than all of the original variables.

In the example of the orange juice data, the correlation matrix (see Table 1.3) brings together the 21 correlation coefficients. It is possible to group the strongly correlated variables into sets, but even for this reduced number of variables, grouping them this way is tedious.

TABLE 1.3
Orange Juice Data: Correlation Matrix

	Odour intensity	Odour typicality	Pulp	Intensity of taste	Acidity	Bitter-ness	Sweet-ness
Odour intensity	1.00	0.58	0.66	−0.27	−0.15	−0.15	0.23
Odour typicality	0.58	1.00	0.77	−0.62	−0.84	−0.88	0.92
Pulp content	0.66	0.77	1.00	−0.02	−0.47	−0.64	0.63
Intensity of taste	−0.27	−0.62	−0.02	1.00	0.73	0.51	−0.57
Acidity	−0.15	−0.84	−0.47	0.73	1.00	0.91	−0.90
Bitterness	−0.15	−0.88	−0.64	0.51	0.91	1.00	−0.98
Sweetness	0.23	0.92	0.63	−0.57	−0.90	−0.98	1.00

1.2.3 Relationships between the Two Studies

The study of individuals and the study of variables are interdependent as they are carried out on the same data table: studying them jointly can only reinforce their respective interpretations.

If the study of individuals led to a distinction between groups of individuals, it is then possible to list the individuals belonging to only one group. However, for high numbers of individuals, it seems more pertinent to characterise them directly by the variables at hand: for example, by specifying that some orange juices are acidic and bitter whereas others have a high pulp content.

Similarly, when there are groups of variables, it may not be easy to interpret the relationships between many variables and we can make use of specific individuals, that is, individuals who are extreme from the point of view of these relationships. In this case, it must be possible to identify the individuals. For example, the link between *acidity-bitterness* can be illustrated by the opposition between two extreme orange juices: *Fresh Pampryl* (orange juice from Spain) versus *Fresh Tropicana* (orange juice from Florida).

1.3 Studying Individuals

1.3.1 The Cloud of Individuals

An individual is a row of the data table, that is, a set of K numerical values. The individuals thus evolve within a space $\mathbb{R}^K$ called "the individual's space." If we endow this space with the usual Euclidean distance, the distance between two individuals i and l is expressed as

$$d(i,l) = \sqrt{\sum_{k=1}^{K}(x_{ik} - x_{lk})^2}.$$

If two individuals have similar values within the table of all K variables, they are also close in the space $\mathbb{R}^K$. Thus, the study of the data table can be conducted geometrically by studying the distances between individuals. We are therefore interested in all of the individuals in $\mathbb{R}^K$, that is, the cloud of individuals (denoted N_I). Analysing the distances between individuals is therefore tantamount to studying the shape of the cloud of points. Figure 1.3 illustrates a cloud of points within a space $\mathbb{R}^K$ for $K = 3$.

FIGURE 1.3
Flight of a flock of starlings illustrating a scatterplot in $\mathbb{R}^K$.

The shape of the cloud N_I remains the same even when translated. The data are also centred, which corresponds to considering $x_{ik} - \bar{x}_k$ rather than x_{ik}. Geometrically, this is tantamount to coinciding the centre of mass of the cloud G_I (with coordinates $\bar{x}_k$ for $k = 1, ..., K$) with the origin of reference (see Figure 1.4). Centring presents technical advantages and is always conducted in PCA.

The operation of reduction (also referred to as standardising), which consists of considering $(x_{ik} - \bar{x}_k)/s_k$ rather than x_{ik}, modifies the shape of the cloud by harmonising its variability in all the directions of the original vectors (i.e., the K variables). Geometrically, it means choosing standard deviation

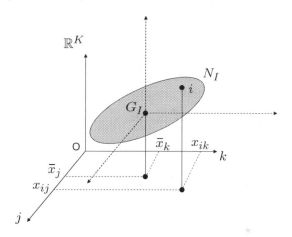

FIGURE 1.4
Scatterplot of the individuals in $\mathbb{R}^K$.

s_k as a unit of measurement in direction k. This operation is essential if the variables are not expressed in the same units. Even when the units of measurement do not differ, this operation is generally preferable as it attaches the same importance to each variable. Therefore, we will assume this to be the case from here on in. Standardised PCA occurs when the variables are centred and reduced, and unstandardised PCA when the variables are only centred. When not otherwise specified, it may be assumed that we are using standardised PCA.

Comment: Weighting Individuals
So far we have assumed that all individuals have the same weight. This applies to almost all applications and is always assumed to be the case. Nevertheless, generalisation with unspecified weights poses no conceptual or practical problems (double weight is equivalent to two identical individuals) and most software packages, including FactoMineR, envisage this possibility (FactoMineR is a package dedicated to Factor Analysis and Data Mining with R; see Section A.2.3 in the appendix). For example, it may be useful to assign a different weight to each individual after having rectified a sample. In all cases, it is convenient to consider that the sum of the weights is equal to 1. If supposed to be of the same weight, each individual will be assigned a weight of $1/I$.

1.3.2 Fitting the Cloud of Individuals

1.3.2.1 Best Plane Representation of N_I

The aim of PCA is to represent the cloud of points in a space with reduced dimensions in an "optimal" manner, that is to say, by distorting the distances between individuals as little as possible. Figure 1.5 gives two representations of three different fruits. The viewpoints chosen for the images of the fruits on the top line make them difficult to identify. On the second row, the fruits can be more easily recognised. What is it which differentiates the views of each fruit between the first and the second lines? In the pictures on the second line, the distances are less distorted and the representations take up more space on the image. The image is a projection of a three-dimensional object in a two-dimensional space.

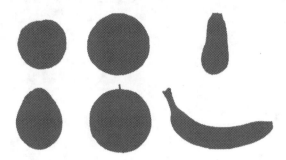

FIGURE 1.5
Two-dimensional representations of fruits: from left to right, an avocado, a melon, and a banana; each row corresponds to a different representation.

For a representation to be successful, it must select an appropriate viewpoint. More generally, PCA means searching for the best representational space (of reduced dimension) thus enabling optimal visualisation of the shape of a cloud with K dimensions. We often use a plane representation alone, which can prove inadequate when dealing with particularly complex data.

To obtain this representation, the cloud N_I is projected on a plane of $\mathbb{R}^K$ denoted P, chosen in such a manner as to minimise distortion of the cloud of points. Plane P is selected so that the distances between the projected points might be as close as possible to the distances between the initial points. Since, in projection, distances can only decrease, we try to make the projected distances as high as possible. By denoting H_i the projection of the individual i on plane P, the problem consists of finding P, with

$$\sum_{i=1}^{I} OH_i^2 \text{ maximum.}$$

The convention for notation uses mechanical terms: O is the centre of gravity, OH_i is a vector, and the criterion is the inertia of the projection of N_I. The criterion which consists of increasing the variance of the projected points to a maximum is perfectly appropriate.

Remark

If the individuals are weighted with different weights p_i, the maximised criterion is $\sum_{i=1}^{I} p_i OH_i^2$.

In some rare cases, it might be interesting to search for the best axial representation of cloud N_I alone. This best axis is obtained in the same way: find the component u_1 when $\sum_{i=1}^{I} OH_i^2$ are maximum (where H_i is the projection of i on u_1). It can be shown that plane P contains component u_1 (the "best" plane contains the "best" component): in this case, these representations are said to be nested. An illustration of this property is presented in Figure 1.6. Planets, which are in a three-dimensional space, are traditionally represented on a component. This component determines their positions as well as possible in terms of their distances from one other (in terms of inertia of the projected cloud). We can also represent planets on a plane according to the same principle: to maximise the inertia of the projected scatterplot (on the plane). This best plane representation also contains the best axial representation.

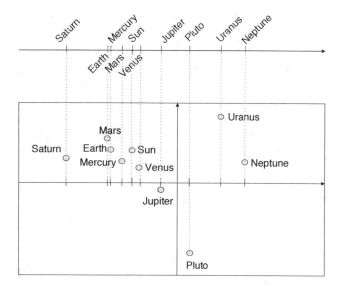

FIGURE 1.6
The best axial representation is nested in the best plane representation of the solar system (18 February 2008).

We define plane P by two nonlinear vectors chosen as follows: vector u_1, which defines the best axis (and which is included in P), and vector u_2 of the plane P orthogonal to u_1. Vector u_2 corresponds to the vector which expresses the greatest variability of N_I once that which is expressed by u_1 is removed. In other words, the variability expressed by u_2 is the best coupling and is independent of that expressed by u_1.

1.3.2.2 Sequence of Axes for Representing N_I

More generally, let us look for nested subspaces of dimensions $s = 1$ to S so that each subspace is of maximum inertia for the given dimension s. The subspace of dimension s is obtained by maximising $\sum_{i=1}^{I} (OH_i)^2$ (where H_i is the projection of i in the subspace of dimension s). As the subspaces are nested, it is possible to choose vector u_s as the vector of the orthogonal subspace for all of the vectors u_t (with $1 \leq t < s$) which define the smaller subspaces.

The first plane (defined by u_1, u_2), i.e., the plane of best representation, is often sufficient for visualising cloud N_I. When S is greater than or equal to 3, we may need to visualise cloud N_I in the subspace of dimension S by using a number of plane representations: the representation on (u_1, u_2) but also that on (u_3, u_4) which is the most complementary to that on (u_1, u_2). However, in certain situations, we might choose to associate (u_2, u_3), for example, in order to highlight a particular phenomenon which appears on these two components.

1.3.2.3 How Are the Components Obtained?

Components in PCA are obtained through diagonalisation of the correlation matrix which extracts the associated eigenvectors and eigenvalues. The eigenvectors correspond to vectors u_s which are each associated with the eigenvalues of rank s (denoted λ_s), as the eigenvalues are ranked in descending order. The eigenvalue λ_s is interpreted as the inertia of cloud N_I projected on the component of rank s or, in other words, the "explained variance" for the component of rank s.

If all of the eigenvectors are calculated ($S = K$), the PCA recreates a basis for the space $\mathbb{R}^K$. In this sense, PCA can be seen as a change of basis in which the first vectors of the new basis play an important role.

Remark
When variables are centred but not standardised, the matrix to be diagonalised is the variance–covariance matrix.

1.3.2.4 Example

The distance between two orange juices is calculated using their seven sensory descriptors. We decided to standardise the data to attribute each descriptor equal influence. Figure 1.7 is obtained from the first two components of the PCA and corresponds to the best plane for representing the cloud of individuals in terms of projected inertia. The inertia projected on the plane is the sum of two eigenvalues, that is, 86.82% ($= 67.77\% + 19.05\%$) of the total inertia of the cloud of points.

The first principal component, that is, the principal axis of variability

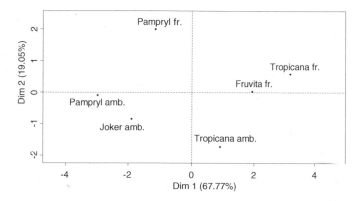

FIGURE 1.7
Orange juice data: plane representation of the scatterplot of individuals.

between the orange juices, separates the two orange juices *Tropicana fr.* and *Pampryl amb.* According to data Table 1.2, we can see that these orange juices are the most extreme in terms of the descriptors *odour typicality* and *bitterness*: *Tropicana fr.* is the most typical and the least bitter while *Pampryl amb.* is the least typical and the most bitter. The second component, that is, the property that separates the orange juices most significantly once the main principal component of variability has been removed, identifies *Tropicana amb.*, which is the least intense in terms of odour, and *Pampryl fr.*, which is among the most intense (see Table 1.2).

Reading this data is tedious when there are a high number of individuals and variables. For practical purposes, we will facilitate the characterisation of the principal components by using the variables more directly.

1.3.3 Representation of the Variables as an Aid for Interpreting the Cloud of Individuals

Let F_s denote the coordinate of the I individuals on component s and $F_s(i)$ its value for individual i. Vector F_s is also called the principal component of rank s. F_s is of dimension I and thus can be considered as a variable. To interpret the relative positions of the individuals on the component of rank s, it may be interesting to calculate the correlation coefficient between vector F_s and the initial variables. Thus, when the correlation coefficient between F_s and a variable k is positive (or indeed negative), an individual with a positive coordinate on component F_s will generally have a high (or low, respectively) value (relative to the average) for variable k.

In the example, F_1 is strongly positively correlated with the variables *odour typicality* and *sweetness* and strongly negatively correlated with the variables *bitter* and *acidic* (see Table 1.4). Thus *Tropicana fr.*, which has the highest coordinate on component 1, has high values for *odour typicality* and *sweetness* and low values for the variables *acidic* and *bitter*. Similarly, we can examine the correlations between F_2 and the variables. It may be noted that the correlations are generally lower (in absolute value) than those with the first principal component. We will see that this is directly linked to the percentage of inertia associated with F_2 which is, by construction, lower than that associated with F_1. The second component can be characterised by the variables *odour intensity* and *pulp content* (see Table 1.4).

TABLE 1.4
Orange Juice Data: Correlation between
Variables and First Two Components

	F_1	F_2
Odour intensity	0.46	0.75
Odour typicality	0.99	0.13
Pulp content	0.72	0.62
Intensity of taste	−0.65	0.43
Acidity	−0.91	0.35
Bitterness	−0.93	0.19
Sweetness	0.95	−0.16

To make these results easier to interpret, particularly in cases with a high number of variables, it is possible to represent each variable on a graph, using its correlation coefficients with F_1 and F_2 as coordinates (see Figure 1.8).

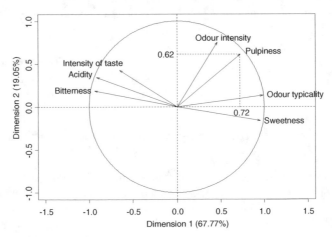

FIGURE 1.8
Orange juice data: visualisation of the correlation coefficients between variables and the principal components F_1 and F_2.

We can now interpret the joint representation of the cloud of individuals with this representation of the variables.

Remark

A variable is always represented within a circle of radius 1 (circle represented in Figure 1.8): indeed, it must be noted that F_1 and F_2 are orthogonal (in the sense that their correlation coefficient is equal to 0) and that a variable cannot be strongly related to two orthogonal components simultaneously. In the following section we shall examine why the variable will always be found within the circle of radius 1.

1.4 Studying Variables

1.4.1 The Cloud of Variables

Let us now consider the data table as a set of columns. A variable is one of the columns in the table, that is, a set of I numerical values, which is represented by a point of the vector space with I dimensions, denoted $\mathbb{R}^I$ (and known as the "variables' space"). The vector connects the origin of $\mathbb{R}^I$ to the point. All of these vectors constitute the cloud of variables and this ensemble is denoted N_K (see Figure 1.9).

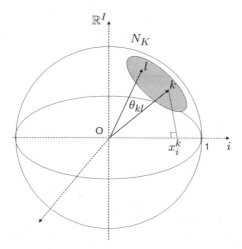

FIGURE 1.9

The scatterplot of the variables N_K in $\mathbb{R}^I$. In the case of a standardised PCA, the variables k are located within a hypersphere of radius 1.

The scalar product between two variables k and l is expressed as

$$\sum_{i=1}^{I} x_{ik} \times x_{il} = \|k\| \times \|l\| \times \cos(\theta_{kl})$$

with $\|k\|$ and $\|l\|$ the norm for variable k and l, and θ_{kl} the angle produced by the vectors representing variables k and l. Since the variables used here are centred, the norm for one variable is equal to its standard deviation multiplied by the square root of I, and the scalar product is expressed as follows:

$$\sum_{i=1}^{I}(x_{ik} - \bar{x}_k) \times (x_{il} - \bar{x}_l) = I \times s_k \times s_l \times \cos(\theta_{kl}).$$

On the right-hand side of the equation, we can identify covariance between variables k and l.

Similarly, by dividing each term in the equation by the standard deviations s_k and s_l of variables k and l, we obtain the following relationship:

$$r(k,l) = \cos(\theta_{kl}).$$

This property is essential in PCA as it provides a geometric interpretation of the correlation. Therefore, in the same way as the representation of cloud N_I can be used to visualise the variability between individuals, a representation of the cloud N_K can be used to visualise all of the correlations (through the angles between variables) or, in other words, the correlation matrix. To facilitate visualisation of the angles between variables, the variables are represented by vectors rather than points.

Generally speaking, the variables being centred and reduced (scaled to unit variance) have a length with a value of 1 (hence the term "standardised variable"). The vector extremities are therefore on the sphere of radius 1 (also called "hypersphere" to highlight the fact that, in general, $I > 3$), as shown in Figure 1.9.

Comment about the Centring
In $\mathbb{R}^K$, when the variables are centred, the origin of the axes is translated onto the mean point. This property is not true for N_K.

1.4.2 Fitting the Cloud of Variables

As is the case for the individuals, the cloud of variables N_K is situated in a space $\mathbb{R}^I$ with a great number of dimensions and it is impossible to visualise the cloud in the overall space. The cloud of variables must therefore be adjusted using the same strategy as for the cloud of individuals. We maximise an equivalent criterion $\sum_{k=1}^{K}(OH_k)^2$ with H_k, the projection of variable k on the subspace with reduced dimensions. Here too, the subspaces are nested and we can identify a series of orthogonal axes S which define the subspaces for dimensions $s = 1$ to S. Vector v_s therefore belongs to a given subspace and is orthogonal to the vectors v_t which make up the smaller subspaces. It can therefore be shown that the vector v_s maximises $\sum_{k=1}^{K}(OH_k^s)^2$ where H_k^s is the projection of variable k on v_s.

Remark

In the individual space $\mathbb{R}^K$, centring the variables causes the origin of the axes to shift to a mean point: the maximised criterion is therefore interpreted as a variance; the projected points must be as dispersed as possible. In $\mathbb{R}^I$, centring has a different effect, as the origin is not the same as the mean point. The projected points should be as far as possible from the origin (although not necessarily dispersed), even if that means being grouped together or merged. This indicates that the position of the cloud N_K with respect to the origin is important.

Vectors v_s $(s = 1, ..., S)$ belong to the space $\mathbb{R}^I$ and consequently can be considered new variables. The correlation coefficient $r(k, v_s)$ between variables k and v_s is equal to the cosine of the angle θ_k^s between Ok and v_s when variable k is centred and scaled, and thus standardised. The plane representation constructed by (v_1, v_2) is therefore pleasing as the coordinates of a variable k correspond to the cosine of the angle θ_k^1 and that of angle θ_k^2, and thus the correlation coefficients between variables k and v_1, and between variables k and v_2. In a plane representation such as this, we can therefore immediately visualise whether or not a variable k is related to a dimension of variability (see Figure 1.10).

By their very construction, variables v_s maximise criterion $\sum_{k=1}^{K}(OH_k^s)^2$. Since the projection of a variable k is equal to the cosine of angle θ_k^s, the criterion maximises

$$\sum_{k=1}^{K} \cos^2 \theta_k^s = \sum_{k=1}^{K} r^2(k, v_s).$$

The above expression illustrates that v_s is the new variable which is the most strongly correlated with all of the initial variables K (with the orthogonality constraint of v_t already found). As a result, v_s can be said to be a synthetic variable. Here, we are experiencing the second aspect of the study of variables (see Section 1.2.2).

Remark

When a variable is not standardised, its length is equal to its standard deviation. In an unstandardised PCA, the criterion can be expressed as follows:

$$\sum_{k=1}^{K}(OH_k^s)^2 = \sum_{k=1}^{K} s_k^2 r^2(k, v_s).$$

This highlights the fact that, in the case of an unstandardised PCA, each variable k is assigned a weight equal to its variance s_k^2.

It can be shown that the axes of representation N_K are in fact eigenvectors of the scalar products matrix between individuals. This property is, in practice, only used when the number of variables exceeds the number of individuals. We will see in the following that these eigenvectors are deducted from those of the correlation matrix.

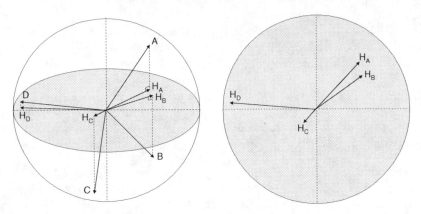

FIGURE 1.10
Projection of the scatterplot of the variables on the main plane of variability. On the left: visualisation in space $\mathbb{R}^I$; on the right: visualisation of the projections in the principal plane.

The best plane representation of the cloud of variables corresponds exactly to the graph representing the variables obtained as an aid to interpreting the representation of individuals (see Figure 1.8). This remarkable property is not specific to the example but applies when carrying out any standardised PCA. This point will be developed further in the following section.

1.5   Relationships between the Two Representations N_I and N_K

So far we have looked for representations of N_I and N_K according to the same principle and from one single data table. It therefore seems natural for these two analyses (N_I in $\mathbb{R}^K$ and N_K in $\mathbb{R}^I$) to be related.

The relationships between the two clouds N_I and N_K are brought together under the general term of "relations of duality." This term refers to the dual approach of one single data table, by considering either the lines or the columns. This approach is also defined by "transition relations" (calculating the coordinates in one space from those in the other). Where $F_s(i)$ is the coordinate of individual i and $G_s(k)$ the coordinate of variable k of the component of rank s, we obtain the following equations:

$$F_s(i) = \frac{1}{\sqrt{\lambda_s}} \sum_{k=1}^{K} x_{ik} \, G_s(k),$$

$$G_s(k) = \frac{1}{\sqrt{\lambda_s}} \sum_{i=1}^{I} (1/I)\, x_{ik}\, F_s(i).$$

This result is essential for interpreting the data, and makes PCA a rich and reliable experimental tool. This may be expressed as follows: individuals are on the same side as their corresponding variables with high values, and opposite their corresponding variables with low values. It must be noted that x_{ik} are centred and carry both positive and negative values. This is one of the reasons why individuals can be so far from the variable for which they carry low values. F_s is referred to as the principal component of rank s; λ_s is the variance of F_s and its square root the length of F_s in $\mathbb{R}^I$; v_s is known as the standardised principal component.

The total inertias of both clouds are equal (and equal to K for standardised PCA) and furthermore, when decomposed component by component, they are identical. This property is remarkable: if S dimensions are enough to perfectly represent N_I, the same is true for N_K. In this case, two dimensions are sufficient. If not, why generate a third synthetic variable which would not differentiate the individuals at all?

1.6 Interpreting the Data

1.6.1 Numerical Indicators

1.6.1.1 Percentage of Inertia Associated with a Component

The first indicators that we shall examine are the ratios between the projected inertias and the total inertia. For component s:

$$\frac{\sum_{i=1}^{I} \frac{1}{I} (OH_i^s)^2}{\sum_{i=1}^{I} \frac{1}{I} (Oi)^2} = \frac{\sum_{k=1}^{K} (OH_k^s)^2}{\sum_{k=1}^{K} Ok^2} = \frac{\lambda_s}{\sum_{s=1}^{K} \lambda_s}.$$

In the most usual case, when the PCA is standardised, $\sum_{s=1}^{K} \lambda_s = K$. When multiplied by 100, this indicator gives the percentage of inertia (of N_I in $\mathbb{R}^K$ or of N_K in $\mathbb{R}^I$) expressed by the component of rank s. This can be interpreted in two ways:

1. As a measure of the quality of data representation; in the example, we say that the first component expresses 67.77% of data variability (see Table 1.5). In a standardised PCA (where $I > K$), we often compare λ_s with 1, the value below which the component of rank s, representing less data than a stand-alone variable, is not worthy of interest.

2. As a measure of the relative importance of the components; in the

example, we say that the first component expresses three times more variability than the second; it affects three times more variables but this formulation is truly precise only when each variable is perfectly correlated with a component.

Because of the orthogonality of the axes (both in $\mathbb{R}^K$ and in $\mathbb{R}^I$), these inertia percentages can be added together for several components; in the example, 86.82% of the data are represented by the first two components (67.77% + 19.05% = 86.82%).

TABLE 1.5
Orange Juice Data: Decomposition of Variability
per Component

	Eigenvalue variance	Percentage of of variance	Cumulative percentage
Comp 1	4.74	67.77	67.77
Comp 2	1.33	19.05	86.81
Comp 3	0.82	11.71	98.53
Comp 4	0.08	1.20	99.73
Comp 5	0.02	0.27	100.00

Let us return to Figure 1.5: the pictures of the fruits on the first line correspond approximately to a projection of the fruits on the plane constructed by components 2 and 3 of PCA, whereas the images on the second line correspond to a projection on plane 1-2. This is why the fruits are easier to recognise on the second line: the more variability (i.e., the more information) collected on plane 1-2 when compared to plane 2-3, the easier it is to grasp the overall shape of the cloud. Furthermore, the banana is easier to recognise in plane 1-2 (the second line), as it retrieves greater inertia on plane 1-2. In concrete terms, as the banana is a longer fruit than a melon, this leads to more marked differences in inertia between the components. As the melon is almost spherical, the percentages of inertia associated with each of the three components are around 33% and therefore the inertia retrieved by plane 1-2 is nearly 66% (as is that retrieved by plane 2-3).

1.6.1.2   Quality of Representation of an Individual or Variable

The quality of representation of an individual i on the component s can be measured by the distance between the point within the space and the projection on the component. In reality, it is preferable to calculate the percentage of inertia of the individual i projected on the component s. Therefore, when θ_i^s is the angle between Oi and u_s, we obtain

$$qlt_s(i) = \frac{\text{Projected inertia of } i \text{ on } u_s}{\text{Total inertia of } i} = \cos^2 \theta_i^s.$$

Using Pythagoras' theorem, this indicator is combined for multiple components and is most often calculated for a plane.

The quality of representation of a variable k on the component of rank s is expressed as

$$qlt_s(k) = \frac{\text{Projected inertia of } k \text{ on } v_s}{\text{Total inertia of } k} = \cos^2 \theta_k^s.$$

This last quantity is equal to $r^2(k, v_s)$, which is why the quality of representation of a variable is only very rarely provided by software. The representational quality of a variable in a given plane is obtained directly on the graph by visually evaluating its distance from the circle of radius 1.

1.6.1.3 Detecting Outliers

Analysing the shape of the cloud N_I also means detecting unusual or remarkable individuals. An individual is considered remarkable if it has extreme values for multiple variables. In the cloud N_I, an individual such as this is far from the cloud's centre of gravity, and its remarkable nature can be evaluated from its distance from the centre of the cloud in the overall space $\mathbb{R}^K$.

In the example, none of the orange juices are particularly extreme (see Table 1.6). The two most extreme individuals are *Tropicana fresh* and *Pampryl ambient*.

TABLE 1.6
Orange Juice Data: Distances from the Individuals to the Centre of the Cloud

Pampryl amb.	Tropicana amb.	Fruvita fr.	Joker amb.	Tropicana fr.	Pampryl fr.
3.03	1.98	2.59	2.09	3.51	2.34

1.6.1.4 Contribution of an Individual or Variable to the Construction of a Component

Outliers have an influence on analysis, and it is interesting to know to what extent their influence affects the construction of the components. Furthermore, some individuals can influence the construction of certain components without being remarkable themselves. Detecting those individuals that contribute to the construction of a principal component helps to evaluate the component's stability. It is also interesting to evaluate the contribution of variables in constructing a component (especially in nonstandardised PCA).

To do so, we decompose the inertia of a component individual by individual (or variable by variable). The inertia explained by the individual i on the component s is

$$\frac{(1/I)(OH_i^s)^2}{\lambda_s} \times 100.$$

Distances intervene in the components by their squares, which augments the roles of those individuals at a greater distance from the origin. Outlying individuals are the most extreme on the component, and their contributions are especially useful when the individuals' weights are different.

Remark
These contributions are combined for multiple individuals.

When an individual contributes significantly (i.e., much more than the others) to the construction of a principal component (for example, *Tropicana ambient* and *Pampryl fresh*; for the second component, see Table 1.7), it is not uncommon for the results of a new PCA constructed without this individual to change substantially: the principal components can change and new oppositions between individuals may appear.

TABLE 1.7
Orange Juice Data: Contribution of
Individuals to the Construction of the
Components

	Dim.1	Dim.2
Pampryl amb.	31.29	0.08
Tropicana amb.	2.76	36.77
Fruvita fr.	13.18	0.02
Joker amb.	12.63	8.69
Tropicana fr.	35.66	4.33
Pampryl fr.	4.48	50.10

Similarly, the contribution of variable k to the construction of component s is calculated. An example of this is presented in Table 1.8.

TABLE 1.8
Orange Juice Data: Contribution of Variables to the
Construction of the Components

	Dim.1	Dim.2
Odour intensity	4.45	42.69
Odour typicality	20.47	1.35
Pulp content	10.98	28.52
Taste intensity	8.90	13.80
Acidity	17.56	9.10
Bitterness	18.42	2.65
Sweetness	19.22	1.89

1.6.2   Supplementary Elements

We here define the concept of active and supplementary (or illustrative) elements. By definition, active elements contribute to the construction of the principal components, contrary to supplementary elements. Thus, the inertia of the cloud of individuals is calculated on the basis of active individuals, and similarly, the inertia of the cloud of variables is calculated on the basis of active variables. The supplementary elements make it possible to illustrate the principal components, which is why they are referred to as "illustrative elements." Contrary to the active elements, which must be homogeneous, we can make use of as many illustrative elements as possible.

1.6.2.1 Representing Supplementary Quantitative Variables

By definition, a supplementary quantitative variable plays no role in calculating the distances between individuals. They are represented in the same way as active variables: to assist in interpreting the cloud of individuals (Section 1.3.3). The coordinate of the supplementary variable k' on the component s corresponds to the correlation coefficient between k' and the principal component s (i.e., the variable whose values are the coordinates of the individuals projected on the component of rank s). k' can therefore be represented on the same graph as the active variables.

More formally, the transition formulae can be used to calculate the coordinate of the supplementary variable k' on the component of rank s:

$$G_s(k') = \frac{1}{\sqrt{\lambda_s}} \sum_{i \subset \{\text{active}\}} x_{ik'} F_s(i) = r(k, F_s),$$

where $\{\text{active}\}$ refers to the set of active individuals. This coordinate is calculated from the active individuals alone.

In the example, in addition to the sensory descriptors, there are also physicochemical variables at our disposal (see Table 1.9). However, our stance remains unchanged, namely, to describe the orange juices based on their sensory profiles. This problem can be enriched using the supplementary variables since we can now link sensory dimensions to the physicochemical variables.

TABLE 1.9
Orange Juice Data: Supplementary Variables

	Glucose	Fructose	Saccharose	Sweetening power	pH	Citric acid	Vitamin C
Pampryl amb.	25.32	27.36	36.45	89.95	3.59	0.84	43.44
Tropicana amb.	17.33	20.00	44.15	82.55	3.89	0.67	32.70
Fruvita fr.	23.65	25.65	52.12	102.22	3.85	0.69	37.00
Joker amb.	32.42	34.54	22.92	90.71	3.60	0.95	36.60
Tropicana fr.	22.70	25.32	45.80	94.87	3.82	0.71	39.50
Pampryl fr.	27.16	29.48	38.94	96.51	3.68	0.74	27.00

The correlations circle (Figure 1.11) represents both the active and supplementary variables. The main component of variability opposes the orange juices perceived as acidic/bitter, slightly sweet and somewhat typical with the orange juices perceived as sweet, typical, not very acidic and slightly bitter. The analysis of this sensory perception is reinforced by the variables *pH* and *saccharose*. Indeed, these two variables are positively correlated with the first component and lie on the side of the orange juices perceived as sweet and

slightly acidic (a high pH index indicates low acidity). One also finds the re-
action known as "saccharose inversion" (or hydrolysis): the saccharose breaks
down into glucose and fructose in an acidic environment (the acidic orange
juices thus contain more fructose and glucose, and less saccharose than the
average).

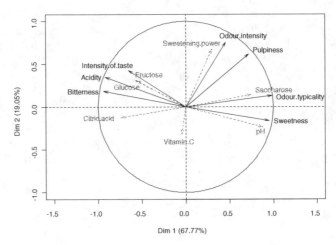

FIGURE 1.11
Orange juice data: representation of the active and supplementary variables.

Remark
When using PCA to explore data prior to a multiple regression, it is advisable
to choose the explanatory variables for the regression model as active variables
for PCA, and to project the variable to be explained (the dependent variable)
as a supplementary variable. This gives some idea of the relationships between
explanatory variables and thus of the need to select explanatory variables.
This also gives us an idea of the quality of the regression: if the dependent
variable is appropriately projected, it will be a well-fitted model.

1.6.2.2   Representing Supplementary Categorical Variables

In PCA, the active variables are quantitative by nature but it is possible to
use information resulting from categorical variables on a purely illustrative
basis (= supplementary), that is, not used to calculate the distances between
individuals.

The categorical variables cannot be represented in the same way as the
supplementary quantitative variables since it is impossible to calculate the
correlation between a categorical variable and F_s. Information about categorical variables lies within their categories. It is quite natural to represent
a categorical variable at the barycentre of all the individuals possessing that
variable. Thus, following projection on the plane defined by the principal

components, these categories remain at the barycentre of the individuals in their plane representation. A categorical variable can thus be regarded as the mean individual obtained from the set of individuals who have it. This is therefore the way in which it will be represented on the graph of individuals.

The information resulting from a supplementary categorical variable can also be represented using a colour code: all of the individuals with the same categorical variable are coloured in the same way. This facilitates visualisation of dispersion around the barycentres associated with specific categories.

In the example, we can introduce the variable *way of preserving* which has two categories *ambient* and *fresh* as well as the variable *origin* of the fruit juice which has also two categories *Florida* and *Other* (see Table 1.10). It seems that sensory perception of the products differs according to their packaging (despite the fact that they were all tasted at the same temperature). The second bisectrix separates the products purchased in the chilled section of the supermarket from the others (see Figure 1.12).

TABLE 1.10
Orange Juice Data: Supplementary
Categorical Variables

	Way of preserving	Origin
Pampryl amb.	Ambient	Other
Tropicana amb.	Ambient	Florida
Fruvita fr.	Fresh	Florida
Joker amb.	Ambient	Other
Tropicana fr.	Fresh	Florida
Pampryl fr.	Fresh	Other

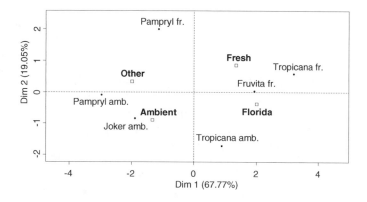

FIGURE 1.12
Orange juice data: plane representation of the scatterplot of individuals with a supplementary categorical variable.

1.6.2.3   Representing Supplementary Individuals

Just as for the variables, we can use the transition formula to calculate the coordinate of a supplementary individual i' on the component of rank s:

$$F_s(i') = \frac{1}{\sqrt{\lambda_s}} \sum_{k=1}^{K} x_{i'k} G_s(k).$$

Note that centring and standardising (if any) are conducted with respect to the averages and the standard deviations calculated from the active individuals only. Moreover, the coordinate of i' is calculated from the active variables alone. Therefore, it is not necessary to have the values of the supplementary individuals for the supplementary variables.

Comment
A supplementary categorical variable can be regarded as a supplementary individual which, for each active variable, would take the average calculated from all of the individuals with this categorical variable.

1.6.3   Automatic Description of the Components

The components provided by the principal component method can be described automatically by all of the variables, whether quantitative or categorical, supplementary or active.

For a quantitative variable, the principle is the same whether the variable is active or supplementary. First, the correlation coefficient between the coordinates of the individuals on the component s and each variable is calculated. We then sort the variables in descending order from the highest coefficient to the weakest and retain the variables with the highest correlation coefficients (absolute values).

Comment
Let us recall that principal components are linear combinations of the active variables, as are synthetic variables. Testing the significance of the correlation coefficient between a component and a variable is thus a biased procedure by its very construction. However, it is useful to sort and select the active variables in this manner to describe the components. On the other hand, for the supplementary variables, the test described corresponds to that traditionally used to test the significance of the correlation coefficient between two variables.

For a categorical variable, we conduct a one-way analysis of variance where we seek to explain the coordinates of the individuals (on the component of rank s) by the categorical variable; we use the sum to zero contrasts $\sum_i \alpha_i = 0$. Then, for each categorical variable, a Student t-test is conducted to compare the average of the individuals who possess that category with the general average (we test $\alpha_i = 0$ considering that the variances of the coordinates are equal for each category). The correlation coefficients are sorted according to the *p*-values in descending order for the positive coefficients and in ascending order for the negative coefficients.

These tips for interpreting such data are particularly useful for understanding those dimensions with a high number of variables.

The data used is made up of few variables. We shall nonetheless give the outputs of the automatic description procedure for the first component as an example. The variables which best characterise component 1 are odour typicality, sweetness, bitterness, and acidity (see Table 1.11).

TABLE 1.11
Orange Juice Data: Description of the First Dimension by the Quantitative Variables

	Correlation	*p*-value
Odour typicality	0.9854	0.0003
Sweetness	0.9549	0.0030
pH	0.8797	0.0208
Aciditiy	−0.9127	0.0111
Bitterness	−0.9348	0.0062

The first component is also characterised by the categorical variable *Origin* as the correlation is significantly different from 0 (*p*-value = 0.00941; see the result in the object `quali` in Table 1.12); the coordinates of the orange juices from Florida are significantly higher than average on the first component, whereas the coordinates of the other orange juices are lower than average (see the results in the object `category` in Table 1.12).

TABLE 1.12
Orange Juice Data: Description of the First Dimension by the Categorical Variables and the Categories of These Categorical Variables

$Dim.1$quali		
	R2	*p*-value
Origin	0.8458	0.0094

$Dim.1$category		
	Estimate	*p*-value
Florida	2.0031	0.0094
Other	−2.0031	0.0094

1.7   Implementation with **FactoMineR**

In this section, we will explain how to carry out a PCA with FactoMineR and how to find the results obtained with the orange juice data. First, load

FactoMineR and then import the data, specifying that the names of the individuals should appear in the first column (`row.names=1`):

```
> library(FactoMineR)
> orange <- read.table("http://factominer.free.fr/bookV2/orange.csv",
    header=TRUE,sep=";",dec=".",row.names=1)
> summary(orange)
```

The PCA is obtained by specifying that, here, variables 8 to 15 are quantitative supplementary whereas variables 16 and 17 are categorical supplementary:

```
> res.pca <- PCA(orange,quanti.sup=8:15,quali.sup=16:17)
```

This command executes the PCA and produces the graph of variables (with the active and supplementary variables, see Figure 1.11) and the graph of individuals (with the individuals and categories of the supplementary categorical variables, see Figure 1.12). To produce a graph with individuals only (see Figure 1.7), we use the function **plot.PCA**:

```
> plot(res.pca,invisible="quali")
```

All the numerical outputs of the PCA function can be summarised by the **summary.PCA** function (this function can simply be applied using the instruction **summary**):

```
> summary(res.pca)
```

Tables 1.4, 1.5, 1.6, 1.7 and 1.8 are obtained using the following lines of code:

```
> round(res.pca$var$coord[,1:2],2)
> round(res.pca$eig,2)
> round(res.pca$ind$dist,2)
> round(res.pca$ind$contrib[,1:2],2)
> round(res.pca$var$contrib[,1:2],2)
```

The command **dimdesc** provides the automatic dimension description from the quantitative (see Table 1.11) and categorical variables (see Table 1.12). The function **lapply** is only used to round (with the function **round**) all of the terms in a list (here within another list):

```
> lapply(dimdesc(res.pca),lapply,round,2)
```

1.8   Additional Results

1.8.1   Testing the Significance of the Components

It may be interesting to compare the percentage of inertia associated with a component (or a plane) to the ones obtained with random data tables of the same dimensions. Those data tables are obtained by simulations according to a multinormal distribution with an identity variance–covariance matrix in order to get the distribution of the percentages of inertia under the independence hypothesis. The quantiles of order 0.95 of these percentages are brought together in the appendix in Tables A.1, A.2, A.3, and A.4 and an example is provided in Section 1.9.3.1.

1.8.2   Variables: Loadings versus Correlations

Our approach, in which we examine the correlations between the variables and the principal components, is widely used. However, other points of view complement this approach, as, for example, when looking at loadings. Loadings are interpreted as the coefficients of the linear combination of the initial variables from which the principal components are constructed. From a numerical point of view, the loadings are equal to the coordinates of the variables divided by the square root of the eigenvalue associated with the component. The loadings are the default outputs of the functions **princomp** and **prcomp**.

From this algebraic point of view, supplementary variables cannot be introduced since they do not intervene in the construction of the components and consequently do not intervene in the linear combination.

Additional details. PCA corresponds to a change of basis which makes it possible to change from the initial variables to their linear combinations when the inertia of the projected scatterplot is maximum. Thus, the loadings matrix corresponds to the transition matrix from the old to the new basis. This matrix corresponds to the eigenvectors of the variance–covariance matrix. This can be expressed as

$$F_s(i) = \sum_{k=1}^{K} L_s(k)(x_{ik} - \bar{x}_k)s_k$$

where $L_s(k)$ denotes the coefficient of the linear combination (loading) of variable k on the component of rank s.

1.8.3   Simultaneous Representation: Biplots

A biplot is a graph in which two sets of objects of different forms are represented. When there is a low number of both individuals and variables, it may be interesting to simultaneously represent the cloud of individuals and the cloud of variables in a biplot. However, this superimposed representation is factitious since the two clouds do not occur within the same space (one belongs to $\mathbb{R}^K$ and the other to $\mathbb{R}^I$). We therefore focus exclusively on the interpretation of the directions of the variables in terms of the individuals: an individual is on the side of the variables for which it takes high values. However, distances between individuals are distorted due to a dilation of each component by the inverse of the square root of the eigenvalue with which it is associated. This distortion is all the more important as inertias of the components of representation are very different. Moreover, it is not possible to represent additional quantitative variables. To obtain a simultaneous representation of the clouds, the function **biplot** should be used.

1.8.4　Missing Values

When analysing data, it is relatively common for values to be missing from the data tables. The simplest way to manage these missing values is to replace each one with the average of the variable for which this value is missing. This procedure gives satisfactory results if not too much data is missing.

Aside from this rather crude technique, there are other more sophisticated methods which draw on the structure of the table, and which tend to yield rather more satisfactory results. We shall briefly outline two possible solutions.

Let us consider two strongly correlated variables x and y whilst taking into account all of the variables for both. In the absence of value y for individual i, it is natural to estimate this value from value x for the same individual (for example, using a simple regression). Let us now consider two individuals i and l for which all of the available values are extremely similar. In the absence of value l for variable k, it is natural to estimate this value from the value of i for the same variable k.

The PCA iterative algorithm integrates these ideas in such a way that all of the data can be dealt with. This algorithm functions according to the following principle. An initial phase attributes values to the missing data (such as the mean of the variable for which the data is missing, for example). Then, two phases are alternated: in the first the PCA is constructed from the previously completed table, and in the second, the coordinates of the individuals and variables of PCA are used to reconstitute the data. This algorithm converges rapidly, and upon convergence, the imputed data no longer contributes to the construction of the dimensions for the PCA. In other words, the PCA is constructed from observed data only. In the data reconstitution phase, the number of dimensions which will be used to reconstitute the data must be defined. This choice of number of axes can be made either in advance or by cross-validation.

In PCA, missing data can be dealt with using the package missMDA which is compatible with FactoMineR. In this way, we first estimate the number of axes to reconstitute the data, using the **estim_ncpPCA** function in missMDA, then we impute the data table using the **imputePCA** function in missMDA, and finally a classic PCA can be constructed from this completed table thanks to the **PCA** function in FactoMineR. The lines of code which must be applied in order to construct a PCA on an incomplete dataset donNA are as follows:

```
> library(missMDA)
> nb <- estim_ncpPCA(donNA)          # Choosing the number of dimensions
> Imputed <- imputePCA(donNA, ncp=nb)  # impute missing values
> res.pca <- PCA(Imputed$completeObs) # Perform PCA
```

The iterative PCA algorithm can suffer from overfitting. In practice, overfitting is caused by over-confidence in the relationships between variables which are not as strong as we might believe (due to missing data). This is why the function uses, by default, a regularised version of this iterative PCA algorithm.

1.8.5 Large Datasets

Data tables in certain fields, such as genomics, contain a great deal more variables than individuals (it is common to have tens of rows and thousands of columns). In this case, we recommend to diagonalise the scalar products matrix instead of the correlation matrix as the FactoMineR package does.

When there are both a great number of individuals and a great number of variables, we can call upon iterative algorithms where the number of dimensions to extract is an explicit input of the function.

1.8.6 Varimax Rotation

The practice of rotation of axes is a technique which stems from common and specific factor analysis (another model-based data analysis method), and is sometimes used in PCA.

It is possible to rotate the representation of the cloud of variables obtained by PCA so that the latter is easier to interpret. Many procedures are available, the most well known being founded on the varimax criterion (the procedure is often referred to as the varimax procedure by misuse of language). Varimax rotation is the rotation which maximises the sum of the squares of the loadings. To carry out the varimax procedure in R, the **varimax** function is used. To successfully perform this procedure, the number of selected axes must be predefined (to represent the cloud of variables).

This procedure has the advantage of providing components which are strongly correlated with certain variables and uncorrelated with others. The disadvantage is that it does not provide nested solutions: the first two components of the two-dimensional solution do not correspond to the first two components of the three-dimensional solution. Besides, the solutions do not maximise the projected inertia of the scatterplot and thus are not optimal in that sense.

1.9   Example: The Decathlon Dataset

1.9.1   Data Description — Issues

This dataset contains the results of decathlon events during two athletic meetings which took place one month apart in 2004: the Olympic Games in Athens which took place on 23 and 24 August, and the Decastar 2004 which took place on 25 and 26 September.

For both competitions, the following information is available for each athlete: performance for each of the 10 events, total number of points (for each event, an athlete earns points based on performance; here the sum of points scored), and final ranking (see Table 1.13). The events took place in the following order: 100 metres, long jump, shot put, high jump, 400 metres (first day) and 110 metre hurdles, discus, pole vault, javelin, 1500 metres (second day).

In this table, rows correspond to an athlete's performance profile at the time of an event and each column to a variable describing the athletes' performances during a meeting. There are 12 quantitative variables (the results for the 10 events, the ranking of the athlete, and the total number of points earned) and one categorical variable (the competition in which the athlete took part).

The dataset is available in the package FactoMineR:

```
> library(FactoMineR)
> data(decathlon)
```

We want to obtain a typology of the performance profiles based on the performances for each of the 10 events, such that two performance profiles might be as close as they are similar.

In addition, we want to obtain a review of the relationships between the results for the different events by studying the correlation coefficients between the variables taken pairwise.

These two results (the first related to the individuals, and the second to the variables) will be compared to describe the typology of the individuals based on the variables and vice versa.

We will also be able to relate the typology of the individuals with two quantitative variables which were not used to construct the distances between individuals as well as a categorical variable *competition*.

1.9.2   Analysis Parameters

1.9.2.1   Choice of Active Elements

To obtain a typology of the athletes based on their performances for the 10 decathlon events, such as "two athletes are close as they have similar performance profiles," the distances between two athletes are defined on the basis

TABLE 1.13
Athletes' Performance in the 10 Decathlon Events (Names in Capital
Letters Are Results of the Decastar Meeting)

Name	100m	Long	Shot	High	400m	110m	Disc	Pole	Jave	1500m	Rank	Nb pts	Competition
Sebrle	10.85	7.84	16.36	2.12	48.36	14.05	48.72	5.00	70.52	280.01	1	8893	OG
Clay	10.44	7.96	15.23	2.06	49.19	14.13	50.11	4.90	69.71	282	2	8820	OG
Karpov	10.5	7.81	15.93	2.09	46.81	13.97	51.65	4.60	55.54	278.11	3	8725	OG
Macey	10.89	7.47	15.73	2.15	48.97	14.56	48.34	4.40	58.46	265.42	4	8414	OG
Warners	10.62	7.74	14.48	1.97	47.97	14.01	43.73	4.90	55.39	278.05	5	8343	OG
Zsivoczky	10.91	7.14	15.31	2.12	49.4	14.95	45.62	4.70	63.45	269.54	6	8287	OG
Hernu	10.97	7.19	14.65	2.03	48.73	14.25	44.72	4.80	57.76	264.35	7	8237	OG
Nool	10.8	7.53	14.26	1.88	48.81	14.8	42.05	5.40	61.33	276.33	8	8235	OG
Bernard	10.69	7.48	14.8	2.12	49.13	14.17	44.75	4.40	55.27	276.31	9	8225	OG
Schwarzl	10.98	7.49	14.01	1.94	49.76	14.25	42.43	5.10	56.32	273.56	10	8102	OG
Pogorelov	10.95	7.31	15.1	2.06	50.79	14.21	44.6	5.00	53.45	287.63	11	8084	OG
Schoenbeck	10.9	7.3	14.77	1.88	50.3	14.34	44.41	5.00	60.89	278.82	12	8077	OG
Barras	11.14	6.99	14.91	1.94	49.41	14.37	44.83	4.60	64.55	267.09	13	8067	OG
Smith	10.85	6.81	15.24	1.91	49.27	14.01	49.02	4.20	61.52	272.74	14	8023	OG
Averyanov	10.55	7.34	14.44	1.94	49.72	14.39	39.88	4.80	54.51	271.02	15	8021	OG
Ojaniemi	10.68	7.5	14.97	1.94	49.12	15.01	40.35	4.60	59.26	275.71	16	8006	OG
Smirnov	10.89	7.07	13.88	1.94	49.11	14.77	42.47	4.70	60.88	263.31	17	7993	OG
Qi	11.06	7.34	13.55	1.97	49.65	14.78	45.13	4.50	60.79	272.63	18	7934	OG
Drews	10.87	7.38	13.07	1.88	48.51	14.01	40.11	5.00	51.53	274.21	19	7926	OG
Parkhomenko	11.14	6.61	15.69	2.03	51.04	14.88	41.9	4.80	65.82	277.94	20	7918	OG
Terek	10.92	6.94	15.15	1.94	49.56	15.12	45.62	5.30	50.62	290.36	21	7893	OG
Gomez	11.08	7.26	14.57	1.85	48.61	14.41	40.95	4.40	60.71	269.7	22	7865	OG
Turi	11.08	6.91	13.62	2.03	51.67	14.26	39.83	4.80	59.34	290.01	23	7708	OG
Lorenzo	11.1	7.03	13.22	1.85	49.34	15.38	40.22	4.50	58.36	263.08	24	7592	OG
Karlivans	11.33	7.26	13.3	1.97	50.54	14.98	43.34	4.50	52.92	278.67	25	7583	OG
Korkizoglou	10.86	7.07	14.81	1.94	51.16	14.96	46.07	4.70	53.05	317	26	7573	OG
Uldal	11.23	6.99	13.53	1.85	50.95	15.09	43.01	4.50	60.00	281.7	27	7495	OG
Casarsa	11.36	6.68	14.92	1.94	53.2	15.39	48.66	4.40	58.62	296.12	28	7404	OG
SEBRLE	11.04	7.58	14.83	2.07	49.81	14.69	43.75	5.02	63.19	291.7	1	8217	Dec
CLAY	10.76	7.4	14.26	1.86	49.37	14.05	50.72	4.92	60.15	301.5	2	8122	Dec
KARPOV	11.02	7.3	14.77	2.04	48.37	14.09	48.95	4.92	50.31	300.2	3	8099	Dec
BERNARD	11.02	7.23	14.25	1.92	48.93	14.99	40.87	5.32	62.77	280.1	4	8067	Dec
YURKOV	11.34	7.09	15.19	2.1	50.42	15.31	46.26	4.72	63.44	276.4	5	8036	Dec
WARNERS	11.11	7.6	14.31	1.98	48.68	14.23	41.1	4.92	51.77	278.1	6	8030	Dec
ZSIVOCZKY	11.13	7.3	13.48	2.01	48.62	14.17	45.67	4.42	55.37	268	7	8004	Dec
MCMULLEN	10.83	7.31	13.76	2.13	49.91	14.38	44.41	4.42	56.37	285.1	8	7995	Dec
MARTINEAU	11.64	6.81	14.57	1.95	50.14	14.93	47.6	4.92	52.33	262.1	9	7802	Dec
HERNU	11.37	7.56	14.41	1.86	51.1	15.06	44.99	4.82	57.19	285.1	10	7733	Dec
BARRAS	11.33	6.97	14.09	1.95	49.48	14.48	42.1	4.72	55.4	282	11	7708	Dec
NOOL	11.33	7.27	12.68	1.98	49.2	15.29	37.92	4.62	57.44	266.6	12	7651	Dec
BOURGUIGNON	11.36	6.8	13.46	1.86	51.16	15.67	40.49	5.02	54.68	291.7	13	7313	Dec

of their performances in the 10 events. Thus, only the performance variables are considered active; the other variables (*number of points*, *rank*, and *competition*) are supplementary.

Here, the athletes are all considered active individuals.

1.9.2.2 Should the Variables Be Standardised?

As the 10 performance variables were measured in different units, it is necessary to standardise them for comparison. The role of each variable in calculating the distances between individuals is then balanced from the point of view of their respective standard deviations. Without standardisation, the variable *1500 metres*, with a standard deviation of 11.53, would have 100 times more influence than the variable *high jump* with a standard deviation of 0.09.

1.9.3   Implementation of the Analysis

To perform this analysis, we use the **PCA** function of the FactoMineR package. Its main input parameters are the dataset, whether or not the variables are standardised, the position of the quantitative supplementary variables in the dataset, and the position of the categorical variables in the dataset (supplementary by definition). By default, all of the variables are standardised (scale.unit=TRUE, a parameter that does not need to be defined), and none of the variables are supplementary (quanti.sup=NULL and quali.sup=NULL, in other words, all the variables are both quantitative and active).

In the example, we specify that variables 11 and 12 (*number of points* and *rank*) are quantitative supplementary, and that variable 13 (*competition*) is categorical supplementary:

```
> res.pca <- PCA(decathlon,quanti.sup=11:12,quali.sup=13)
```

The **PCA** function provides the graph of individuals and the graph of variables as well as the outputs that can be summarized by the function **summary.PCA**:

```
> summary(res.pca, nb.dec=2, nbelements=Inf)

Call:
PCA(X = decathlon, quanti.sup = 11:12, quali.sup = 13)

Eigenvalues
                       Dim.1  Dim.2  Dim.3  Dim.4  Dim.5  Dim.6  Dim.7  Dim.8
Variance                3.27   1.74   1.40   1.06   0.68   0.60   0.45   0.40
% of var.              32.72  17.37  14.05  10.57   6.85   5.99   4.51   3.97
Cumulative % of var.   32.72  50.09  64.14  74.71  81.56  87.55  92.06  96.03
                       Dim.9 Dim.10
Variance                0.21   0.18
% of var.               2.15   1.82
Cumulative % of var.   98.18 100.00

Individuals
```

	Dist	Dim.1	ctr	cos2	Dim.2	ctr	cos2	Dim.3	ctr	cos2
SEBRLE	2.37	0.79	0.47	0.11	0.77	0.84	0.11	0.83	1.19	0.12
CLAY	3.51	1.23	1.14	0.12	0.57	0.46	0.03	2.14	7.96	0.37
KARPOV	3.40	1.36	1.38	0.16	0.48	0.33	0.02	1.96	6.64	0.33
BERNARD	2.76	-0.61	0.28	0.05	-0.87	1.07	0.10	0.89	1.37	0.10
YURKOV	3.02	-0.59	0.26	0.04	2.13	6.38	0.50	-1.23	2.61	0.16
WARNERS	2.43	0.36	0.09	0.02	-1.68	3.99	0.48	0.77	1.02	0.10
ZSIVOCZKY	2.56	0.27	0.06	0.01	-1.09	1.68	0.18	-1.28	2.86	0.25
McMULLEN	2.56	0.59	0.26	0.05	0.23	0.07	0.01	-0.42	0.30	0.03
MARTINEAU	3.74	-2.00	2.97	0.28	0.56	0.44	0.02	-0.73	0.93	0.04
HERNU	2.79	-1.55	1.78	0.31	0.49	0.33	0.03	0.84	1.23	0.09
BARRAS	1.95	-1.34	1.34	0.47	-0.31	0.14	0.03	0.00	0.00	0.00
NOOL	3.73	-2.34	4.10	0.39	-1.97	5.43	0.28	-1.34	3.10	0.13
BOURGUIGNON	4.30	-3.98	11.80	0.86	0.20	0.06	0.00	1.33	3.05	0.10
Sebrle	4.84	4.04	12.16	0.70	1.37	2.62	0.08	-0.29	0.15	0.00
Clay	4.65	3.92	11.45	0.71	0.84	0.98	0.03	0.23	0.09	0.00
Karpov	5.01	4.62	15.91	0.85	0.04	0.00	0.00	-0.04	0.00	0.00
Macey	3.43	2.23	3.72	0.42	1.04	1.52	0.09	-1.86	6.03	0.29
Warners	2.98	2.17	3.51	0.53	-1.80	4.57	0.37	0.85	1.26	0.08
Zsivoczky	2.57	0.93	0.64	0.13	1.17	1.92	0.21	-1.48	3.79	0.33
Hernu	1.82	0.89	0.59	0.24	-0.62	0.54	0.11	-0.90	1.40	0.24
Nool	3.10	0.30	0.07	0.01	-1.55	3.35	0.25	1.36	3.19	0.19
Bernard	2.83	1.91	2.71	0.45	-0.09	0.01	0.00	-0.76	1.00	0.07
Schwarzl	1.97	0.08	0.00	0.00	-1.35	2.57	0.47	0.82	1.17	0.17
Pogorelov	2.38	0.54	0.22	0.05	0.77	0.83	0.10	1.35	3.15	0.32
Schoenbeck	1.80	0.11	0.01	0.00	-0.04	0.00	0.00	0.74	0.95	0.17
Barras	2.22	0.00	0.00	0.00	0.36	0.18	0.03	-1.57	4.28	0.50
Smith	3.54	0.87	0.56	0.06	1.06	1.58	0.09	-1.64	4.69	0.22
Averyanov	2.52	0.35	0.09	0.02	-1.56	3.41	0.38	0.28	0.14	0.01
Ojaniemi	2.34	0.38	0.11	0.03	-0.77	0.84	0.11	-0.37	0.24	0.03
Smirnov	2.02	-0.48	0.17	0.06	-1.06	1.58	0.28	-1.23	2.62	0.37
Qi	1.76	-0.43	0.14	0.06	-0.33	0.15	0.03	-1.07	1.99	0.37
Drews	3.42	-0.25	0.05	0.01	-3.08	13.33	0.81	1.05	1.93	0.09
Parkhomenko	3.49	-1.07	0.85	0.09	2.09	6.15	0.36	-1.00	1.74	0.08
Terek	3.28	-0.68	0.35	0.04	0.54	0.40	0.03	2.21	8.47	0.45
Gomez	2.61	-0.29	0.06	0.01	-1.20	2.01	0.21	-1.31	2.96	0.25
Turi	3.07	-1.54	1.77	0.25	0.43	0.26	0.02	0.51	0.46	0.03
Lorenzo	3.51	-2.41	4.32	0.47	-1.58	3.52	0.20	-1.50	3.92	0.18
Karlivans	2.70	-1.99	2.97	0.54	-0.29	0.12	0.01	-0.34	0.20	0.02
Korkizoglou	3.98	-0.96	0.68	0.06	2.07	6.00	0.27	2.59	11.61	0.42
Uldal	2.95	-2.56	4.89	0.76	0.25	0.08	0.01	-0.42	0.30	0.02
Casarsa	4.92	-2.86	6.09	0.34	3.80	20.25	0.60	0.03	0.00	0.00

Variables

	Dim.1	ctr	cos2	Dim.2	ctr	cos2	Dim.3	ctr	cos2
100m	-0.77	18.34	0.60	0.19	2.02	0.04	-0.18	2.42	0.03
Long.jump	0.74	16.82	0.55	-0.35	6.87	0.12	0.18	2.36	0.03
Shot.put	0.62	11.84	0.39	0.60	20.61	0.36	-0.02	0.04	0.00
High.jump	0.57	10.00	0.33	0.35	7.06	0.12	-0.26	4.79	0.07
400m	-0.68	14.12	0.46	0.57	18.67	0.32	0.13	1.23	0.02
110m.hurdle	-0.75	17.02	0.56	0.23	3.01	0.05	-0.09	0.61	0.01
Discus	0.55	9.33	0.31	0.61	21.16	0.37	0.04	0.13	0.00
Pole.vault	0.05	0.08	0.00	-0.18	1.87	0.03	0.69	34.06	0.48
Javeline	0.28	2.35	0.08	0.32	5.78	0.10	-0.39	10.81	0.15
1500m	-0.06	0.10	0.00	0.47	12.95	0.22	0.78	43.54	0.61

Supplementary continuous variables

	Dim.1	cos2	Dim.2	cos2	Dim.3	cos2
Rank	| -0.67	0.45 |	0.05	0.00 |	-0.06	0.00 |
Points	| 0.96	0.91 |	-0.02	0.00 |	-0.07	0.00 |

Supplementary categories

	Dist	Dim.1	cos2	v.test	Dim.2	cos2	v.test	Dim.3	cos2	v.test
Decastar	|0.95 |	-0.60	0.40	-1.43|	-0.04	0.00	-0.12|	0.29	0.09	1.05|
OlympicG	|0.44 |	0.28	0.40	1.43|	0.02	0.00	0.12|	-0.13	0.09	-1.05|

1.9.3.1 Choosing the Number of Dimensions to Examine

Studying the inertia of the principal components allows us, on the one hand, to see whether the variables are structured (presence of correlations between variables) and, in addition, to determine the number of components to be interpreted.

The results in `res.pca$eig` correspond to the eigenvalue (i.e., the inertia or the variance explained) associated with each of the components; the percentage of inertia associated with each component and the cumulative sum of these percentages. These values can be plotted on a bar chart (see Figure 1.13):

```
> barplot(res.pca$eig[,1],main="Eigenvalues",
+    names.arg=paste("dim",1:nrow(res.pca$eig)))
```

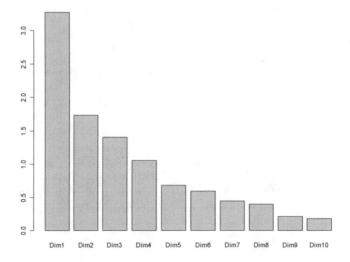

FIGURE 1.13
Decathlon data: eigenvalues associated with each dimension provided by the PCA.

The first two main factors of variability summarise 50% of the total inertia, in other words, 50% of the total variability of the cloud of individuals (or variables) is represented by the plane. The importance of this percentage may not be evaluated without taking into account the number of active individuals and active variables. It may be interesting to compare this percentage with the 0.95-quantile of the distribution of the percentages obtained by simulating data tables of equivalent size on the basis of a normal distribution. According to Table A.3 in the appendix, this quantile obtained for 40 individuals and 10 variables is worth 38%: even if the percentage of 50% seems relatively low, it expresses a significant structure in the data.

However, the variability of performances cannot be summarised by the first two dimensions alone. It may also be interesting to interpret components 3 and 4 for which inertia is greater than 1 (this value is used as a reference because it represents, in the case of standardised variables, the contribution of a single variable).

1.9.3.2 Studying the Cloud of Individuals

The cloud of individuals representation (Figure 1.14) is a default output of the **PCA** function. The supplementary categorical variables are also represented on the graph of individuals through their categories.

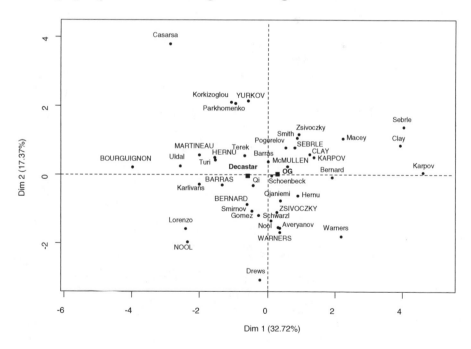

FIGURE 1.14
Decathlon data: graph of individuals (in capitals the results for the Decastar meeting).

Bourguignon and *Karpov* have very different performance profiles since they are opposed according to the main axis of variability. *Casarsa* seems to be an unusual athlete in that his results are extreme for both the first and second principal components.

The results of these individuals can be found in the outputs of the function **summary.PCA** (cf. 1.9.3) or in the object `res.pca$ind`.

The results of the supplementary categorical variables can also be found in the outputs of the function **summary.PCA** (cf. 1.9.3) or in the object `res.pca$quali.sup`. The table contains the coordinates, cosine squared, and v-tests for each category (which indicates whether a category characterises a component or not: here we find the v-test presented in Section 3.7.2.2).

We can also build the graph of individuals with components 3 and 4 (see Figure 1.15). To do so, we use the function **plot.PCA** (which may be used with both **plot** or **plot.PCA**). Note that both the type of objects to be represented (`choix="ind"`) and the axes of representation (`axes = 3:4`) must be specified:

```
> plot(res.pca,choix="ind",axes=3:4)
```

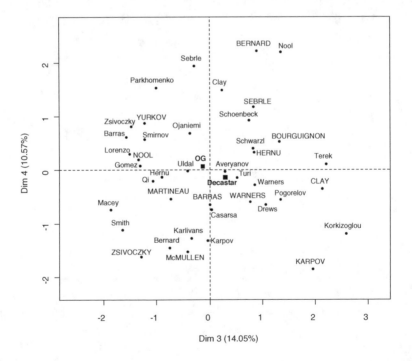

FIGURE 1.15
Decathlon data: graph of individuals on the plane 3-4 (in capitals the results of the Decastar meeting).

Further Details. Individuals can be colour-coded according to categorical variables. In the `plot` function, we specify that the individuals are coloured according to the 13th variable (`habillage=13`). It is also possible to change the font size with the argument `cex` (`"cex = 0.7"` instead of 1 by default).

```
> plot(res.pca,choix="ind",habillage=13,cex=0.7)
```

Confidence ellipses can be drawn around the categories of a supplementary categorical variable (i.e., around the barycentre of the individuals characterised by the category). These ellipses are well adapted to plane representation and enable us to visualise whether or not two categories differ significantly. For this, we consider that the data are multinormally distibuted (which is reasonable when the data are sufficiently numerous as we are interested in centres of gravity and therefore averages).

In practice, the confidence ellipses (see Figure 1.16) are drawn with

```
> plotellipses(res.pca,cex=0.8)
```

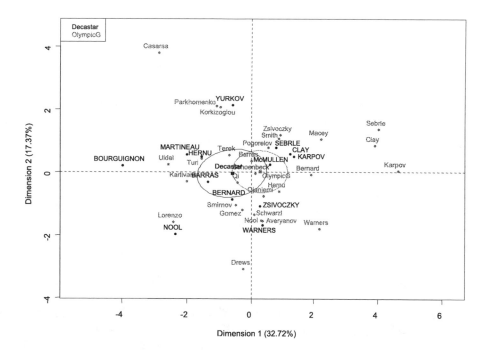

FIGURE 1.16
Decathlon data: confidence ellipses around the categories on the main plane induced by components 1 and 2.

1.9.3.3   Studying the Cloud of Variables

This representation of the cloud of variables enables rapid visualisation of the positive or negative links between variables, the presence of groups of variables that are closely related, and so forth.

The representation of the cloud of variables (Figure 1.17) is also a default output of the **PCA** function. The active variables appear as solid lines while the supplementary variables appear dashed.

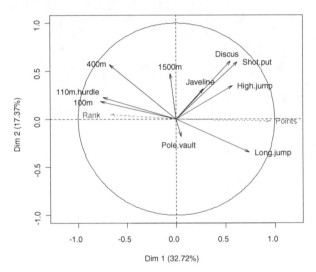

FIGURE 1.17
Decathlon data: graph of variables.

The variables *100m* and *long jump* are negatively correlated: therefore, an athlete who runs 100 metres quickly will generally jump a long way.

The variables *100m*, *400m*, and *110m hurdles* are positively correlated, that is, some athletes perform well in all four events while others do not.

Overall, the variables relating to speed are negatively correlated with the first principal component while the variables *shot put* and *long jump* are positively correlated with this component. The coordinates of these active variables can be found in the outputs of the function **summary.PCA** (cf. 1.9.3), which gives the coordinates, the representation quality of the variables (cosine squared), and their contributions to the construction of the components.

The coordinates (and representation quality) of the supplementary variables can be found as well in the outputs of the function **summary.PCA** (cf. 1.9.3).

The graph of variables on components 3 and 4 (see Figure 1.18) can also be obtained as follows:

```
> plot(res.pca,choix="var",axes=3:4)
```

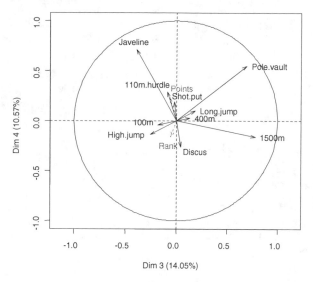

FIGURE 1.18

Decathlon data: graph of variables on the plane induced by components 3-4.

It must be noted that it is possible to obtain an automatic description of the main components using the `dimdesc` function (see Section 1.6.3):

```
> dimdesc(res.pca)
$Dim.1
$Dim.1$quanti
                correlation       P-value
Points           0.9561543 2.099191e-22
Long.jump        0.7418997 2.849886e-08
Shot.put         0.6225026 1.388321e-05
High.jump        0.5719453 9.362285e-05
Discus           0.5524665 1.802220e-04
Rank            -0.6705104 1.616348e-06
400m            -0.6796099 1.028175e-06
110m.hurdle     -0.7462453 2.136962e-08
100m            -0.7747198 2.778467e-09


$Dim.2
$Dim.2$quanti
              correlation       P-value
Discus         0.6063134 2.650745e-05
Shot.put       0.5983033 3.603567e-05
400m           0.5694378 1.020941e-04
1500m          0.4742238 1.734405e-03
High.jump      0.3502936 2.475025e-02
Javelin        0.3169891 4.344974e-02
Long.jump     -0.3454213 2.696969e-02
```

This function is very useful when there are a great number of variables. We see here that the first component is mainly due to the variable *number of points* (with a correlation coefficient of 0.96) and the variable *100m* (with a negative correlation of −0.77). The second component is described by two quantitative variables only (*discus* and *shot put*). No category of any categorical variable characterises components 1 and 2 with a confidence level of 95%. The level of confidence can be changed (`proba = 0.2`, instead of `proba = 0.05` by default), which provides for the quantitative variables:

```
> dimdesc(res.pca,proba=0.2)
$Dim.1$quali
            P-value
Competition 0.1552515

$Dim.1$category
          Estimate   P-value
OlympicG   0.4393744 0.1552515
Decastar  -0.4393744 0.1552515
```

With this confidence level, we can say that both of the two categories *Olympic Games* and *Decastar* have coordinates that are significantly different from 0 on the first component. As the value is positive (negative) for the Olympic Games (Decastar) we can say that individuals who participated in the Olympic Games tended to have positive coordinates (or negative, respectively) on component 1.

1.9.3.4   Joint Analysis of the Cloud of Individuals and the Cloud of Variables

The representations of both the cloud of individuals and the cloud of variables are to be analysed together. In other words, differences between individuals can be explained by the variables, and relationships between variables can be illustrated by individuals.

On the whole, the first component opposes performance profiles that are "uniformly high" (i.e., athletes that are good in all events) such as *Karpov* at the Olympics to performance profiles that are (relatively!) "weak in all events" such as *Bourguignon* at the Decastar meeting.

Furthermore, the first component is mainly linked to the events using qualities relating to a burst of energy (*100m, 400m, 110m hurdles,* and *long jump*). These four variables constitute a relatively homogeneous group: the correlation between any two of these performances is higher than 0.52 (see the correlation matrix on page 43). With just one exception, these variables have the highest coefficients. This group of variables "draws" (i.e., contributes to the construction of) the first principal component and the overall score. It must here be emphasised that the first principal component is the combination that best sums up all the variables. In this example, the automatic summary provided by the PCA corresponds almost exactly with the official summary (the number of points).

The second component opposes the variables of endurance (*400m* and

1500m) and power (*discus, shot put*). Notably, it separates the performance profiles that are considered "weak," which suggests that the best performance profiles are balanced: even among the weakest profiles, athletes can be specialised. Note that power (*discus* and *shot put*) is not correlated with speed (*100m, long jump, 110m hurdles*).

As these two variables are not linearly related, there are powerful and fast individuals (all-round athletes with high values on the first component), powerful individuals who are not so fast (corresponding to high values on component 2), and individuals who are not powerful but fast (with low coordinates on component 2).

This can be illustrated by comparing *Casarsa* and *Lorenzo* from the standardised data (see the table on the next page). *Casarsa* performs well in the events that require power and poorly in speed-related events, while the opposite is true for *Lorenzo*. It must be noted that these two athletes have a low coordinate on the first principal component and therefore do not have good overall performances. Their strengths, which are lessened by the second component, must therefore be relativised compared to their overall performance.

The variable **number of points** seems to be entirely unrelated to this component (correlation of −0.02, see list of coordinates of the supplementary variables).

The third component is mainly related to the 1500 metres and to a lesser extent, to the pole vault. It opposes these two events: athletes that do not perform well in the 1500 metres (N.B. as it is a variable relating to speed, a high value indicates rather poor performance) do however obtain good results in the pole vault (i.e., see standardised values for *Terek*: 1.96 in the pole vault and 0.98 in the 1500 metres).

This third component mainly highlights four individuals that are particularly weak in the 1500 metres: *Clay* and *Karpov* at Decastar (with standardised values of 1.95 and 1.84, respectively) and *Terek* and *Kokhizoglou* at the Olympic Games (with standardised values of 0.98 and 3.29). These four individuals contribute up to 34.7% of the inertia of component 3.

The fourth component is correlated with the variable *javelin* and, to a lesser extent, the variable *pole vault*. Three profiles are characterised by these two events: *Bernard* at the Decastar meeting, and *Sebrle* and *Nool* at the Olympic Games. These three athletes contribute up to 31.3% of the inertia of this component.

It must be noted that the representations of the individuals and variables are only approximate representations of the data table on the one hand, and of the correlation (or variance–covariance) matrix on the other. It is therefore necessary to support the interpretation by referring back to the data. The means and standard deviations by variable, the standardised data, and the correlation matrix can all be obtained using the following lines of code:

```
> res.pca$call$centre
> res.pca$call$ecart.type
      100m  Long  shot  Haut  400m 110mH  Disc  Pole  Jave  1500m
```

```
mean    11.00   7.26 14.48   1.98 49.62 14.61 44.33   4.76 58.32 279.02
sd       0.26   0.31  0.81   0.09  1.14  0.47  3.34   0.27  4.77  11.53
```

Computing the standardised data is useful to facilitate comparison of the data with the average in terms of number of standard deviations, but also to compare one variable's values to another.

```
> round(scale(decathlon[,1:12]),2)
             100m  Long  Shot  High 400m 110mH  Disc  Pole  Jave 1500m  Rank Nbpts
Sebrle      -0.56  1.83  2.28  1.61 -1.09 -1.18  1.30  0.85  2.53  0.08 -1.40  2.59
Clay        -2.12  2.21  0.91  0.94 -0.37 -1.01  1.71  0.49  2.36  0.25 -1.28  2.38
Karpov      -1.89  1.74  1.76  1.27 -2.43 -1.35  2.17 -0.58 -0.58 -0.08 -1.15  2.10
Macey       -0.41  0.66  1.52  1.95 -0.56 -0.10  1.19 -1.30  0.03 -1.17 -1.03  1.19
Warners     -1.44  1.52  0.00 -0.08 -1.43 -1.26 -0.18  0.49 -0.61 -0.08 -0.90  0.99
Zsivoczky   -0.33 -0.38  1.01  1.61 -0.19  0.73  0.38 -0.22  1.06 -0.81 -0.77  0.82
Hernu       -0.11 -0.22  0.21  0.60 -0.77 -0.75  0.12  0.14 -0.12 -1.26 -0.65  0.68
Nool        -0.75  0.85 -0.26 -1.09 -0.70  0.41 -0.67  2.29  0.62 -0.23 -0.52  0.67
Bernard     -1.17  0.70  0.39  1.61 -0.42 -0.92  0.13 -1.30 -0.63 -0.23 -0.39  0.64
Schwarzl    -0.07  0.73 -0.57 -0.41  0.12 -0.75 -0.56  1.21 -0.41 -0.47 -0.27  0.28
Pogorelov   -0.18  0.16  0.76  0.94  1.02 -0.84  0.08  0.85 -1.01  0.74 -0.14  0.23
Schoenbeck  -0.37  0.13  0.36 -1.09  0.59 -0.56  0.02  0.85  0.53 -0.02 -0.02  0.21
Barras       0.54 -0.85  0.53 -0.41 -0.18 -0.50  0.15 -0.58  1.29 -1.02  0.11  0.18
Smith       -0.56 -1.42  0.93 -0.75 -0.30 -1.26  1.39 -2.02  0.66 -0.54  0.24  0.05
Averyanov   -1.70  0.25 -0.04 -0.41  0.09 -0.46 -1.32  0.14 -0.79 -0.69  0.36  0.05
Ojaniemi    -1.21  0.76  0.60 -0.41 -0.43  0.86 -1.18 -0.58  0.20 -0.28  0.49  0.00
Smirnov     -0.41 -0.60 -0.72 -0.41 -0.44  0.35 -0.55 -0.22  0.53 -1.35  0.62 -0.04
Qi           0.24  0.25 -1.12 -0.08  0.03  0.37  0.24 -0.94  0.51 -0.55  0.74 -0.21
Drews       -0.49  0.38 -1.71 -1.09 -0.96 -1.26 -1.25  0.85 -1.41 -0.41  0.87 -0.23
Parkhomenko  0.54 -2.05  1.47  0.60  1.23  0.58 -0.72  0.14  1.55 -0.09  0.99 -0.26
Terek       -0.30 -1.01  0.82 -0.41 -0.05  1.09  0.38  1.93 -1.59  0.97  1.12 -0.33
Gomez        0.31  0.00  0.11 -1.43 -0.87 -0.42 -1.00 -1.30  0.50 -0.80  1.25 -0.41
Turi         0.31 -1.11 -1.04  0.60  1.78 -0.73 -1.33  0.14  0.21  0.94  1.37 -0.87
Lorenzo      0.39 -0.73 -1.52 -1.43 -0.24  1.64 -1.22 -0.94  0.01 -1.37  1.50 -1.21
Karlivans    1.26  0.00 -1.43 -0.08  0.80  0.79 -0.29 -0.94 -1.12 -0.03  1.63 -1.23
Korkizoglou -0.52 -0.60  0.40 -0.41  1.34  0.75  0.52 -0.22 -1.09  3.25  1.75 -1.26
Uldal        0.88 -0.85 -1.15 -1.43  1.16  1.03 -0.39 -0.94  0.35  0.23  1.88 -1.49
Casarsa      1.38 -1.83  0.54 -0.41  3.11  1.66  1.28 -1.30  0.06  1.46  2.01 -1.76
SEBRLE       0.16  1.01  0.43  1.05  0.17  0.18 -0.17  0.93  1.01  1.09 -1.40  0.62
CLAY        -0.91  0.44 -0.26 -1.31 -0.21 -1.18  1.89  0.57  0.38  1.93 -1.28  0.34
KARPOV       0.08  0.13  0.36  0.71 -1.08 -1.09  1.37  0.57 -1.66  1.81 -1.15  0.27
BERNARD      0.08 -0.09 -0.28 -0.64 -0.60  0.81 -1.02  2.01  0.92  0.09 -1.03  0.18
YURKOV       1.30 -0.54  0.86  1.38  0.70  1.49  0.57 -0.15  1.06 -0.22 -0.90  0.09
WARNERS      0.43  1.07 -0.20  0.04 -0.81 -0.80 -0.95  0.57 -1.36 -0.08 -0.77  0.07
ZSIVOCZKY    0.50  0.13 -1.21  0.37 -0.86 -0.92  0.40 -1.23 -0.61 -0.94 -0.65  0.00
McMULLEN    -0.64  0.16 -0.87  1.72  0.25 -0.48  0.02 -1.23 -0.40  0.52 -0.52 -0.03
MARTINEAU    2.44 -1.42  0.11 -0.30  0.45  0.69  0.97  0.57 -1.24 -1.45 -0.39 -0.59
HERNU        1.41  0.95 -0.08 -1.31  1.29  0.96  0.20  0.21 -0.23  0.52 -0.27 -0.80
BARRAS       1.26 -0.92 -0.47 -0.30 -0.12 -0.27 -0.66 -0.15 -0.60  0.25 -0.14 -0.87
NOOL         1.26  0.03 -2.18  0.04 -0.36  1.45 -1.90 -0.51 -0.18 -1.06 -0.02 -1.03
BOURGUIGNON  1.38 -1.45 -1.23 -1.31  1.34  2.26 -1.14  0.93 -0.75  1.09  0.11 -2.02
```

Let us illustrate the standardised data table for the variable *100m*. Individuals with a negative standardised value (positive) take less time (or more, respectively) to run the 100m than the mean of all athletes, which means that they run faster (or slower) than the average.

In addition, the two individuals with the most extreme results are *Clay* and *MARTINE*. *Clay* has a particularly good time since his standardised data is equal to −2.12; in other words, he is at 2.12 standard deviations below the average. *MARTINE* takes a value of 2.44, meaning that his time is particularly long, since he is located at 2.44 standard deviations above the mean: he is particularly slow compared to the average.

If the variable *100m* had followed normal distribution, these two individuals would be part of 5% of the most extreme individuals (95% of the distribution is between -1.96 and 1.96).

The performances of a given individual can be compared for two events. *Clay* is better than the average for the discus and the javelin (the standardised data are positive for both variables), but he is better at the javelin than at the discus (his standardised data is more extreme for the *javelin* than for the *discus*).

The correlation matrix can be obtained with the following code (for covariance matrix, replace **cor** by **cov**):

```
> round(cor(decathlon[,1:12]),2)
        100m  Long  Shot  High  400m 110mH  Disc  Pole  Jave 1500m  Rank  Nbpts
100m    1.00 -0.60 -0.36 -0.25  0.52  0.58 -0.22 -0.08 -0.16 -0.06  0.30 -0.68
Long   -0.60  1.00  0.18  0.29 -0.60 -0.51  0.19  0.20  0.12 -0.03 -0.60  0.73
Shot   -0.36  0.18  1.00  0.49 -0.14 -0.25  0.62  0.06  0.37  0.12 -0.37  0.63
High   -0.25  0.29  0.49  1.00 -0.19 -0.28  0.37 -0.16  0.17 -0.04 -0.49  0.58
400m    0.52 -0.60 -0.14 -0.19  1.00  0.55 -0.12 -0.08  0.00  0.41  0.56 -0.67
110mH   0.58 -0.51 -0.25 -0.28  0.55  1.00 -0.33  0.00  0.01  0.04  0.44 -0.64
Disc   -0.22  0.19  0.62  0.37 -0.12 -0.33  1.00 -0.15  0.16  0.26 -0.39  0.48
Pole   -0.08  0.20  0.06 -0.16 -0.08  0.00 -0.15  1.00 -0.03  0.25 -0.32  0.20
Jave   -0.16  0.12  0.37  0.17  0.00  0.01  0.16 -0.03  1.00 -0.18 -0.21  0.42
1500m  -0.06 -0.03  0.12 -0.04  0.41  0.04  0.26  0.25 -0.18  1.00  0.09 -0.19
Rank    0.30  0.60 -0.37 -0.49  0.56  0.44 -0.39 -0.32 -0.21  0.09  1.00 -0.74
Nbpts  -0.68  0.73  0.63  0.58 -0.67 -0.64  0.48  0.20  0.42 -0.19 -0.74  1.00
```

These correlations between variables can be viewed using the **pairs** function (see Figure 1.19):

```
> pairs(decathlon[,c(1,2,6,10)])
```

1.9.3.5   Comments on the Data

All athletes who participated in both decathlons certainly focused their physical preparation on their performances at the Olympic Games. Indeed, they all performed better at the Olympic Games than at the Decastar meeting. We can see that the dots representing a single athlete (for example, *Sebrle*) are in roughly the same direction. This means, for example, that *Sebrle* is good at the same events for both decathlons, but that the dot corresponding to his performance at the Olympic Games is more extreme, so he obtained more points during the Olympics than at the Decastar meeting.

This data can be interpreted in two different ways:

1. Athletes that participate in the Olympic Games perform better (on average) than those participating in the Decastar meeting.

2. During the Olympics, athletes are more motivated by the challenge, they tend to be fitter, etc.

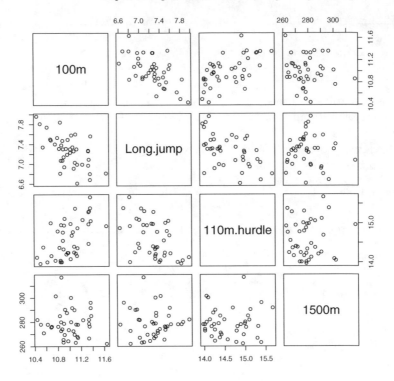

FIGURE 1.19
Decathlon data: cloud of variables *100m, long jump, 110m hurdles,* and *1500m.*

From the point of view of component 2, however, there is no overall difference between the Olympics and Decastar. Overall, athletes' performances may have improved but their profiles have not changed. Only *Zsivoczky* changed from a rather powerful profile to a rather fast profile. His performances in shot put and javelin are worse at Decastar compared to the Olympics, with throws of 15.31m (standardised value of 1.02) and 13.48m (−1.22) for shot put and 63.45m (1.08) and 55.37m (−0.62) for javelin. However, he improved his performance in the 400 metres: 49.40 seconds (−0.19) and then 48.62 seconds (−0.87) and for the 110m hurdles: 14.95 seconds (0.74) and then 14.17 seconds (−0.94).

1.10   Example: The Temperature Dataset

1.10.1   Data Description — Issues

In this example, we investigate the climates of different European countries. To do so, temperatures (in Celsius) were collected monthly for the main European capitals and other major cities. In addition to the monthly temperatures, the average annual temperature and the thermal amplitude (difference between the maximum monthly average and the minimum monthly average of a city) were recorded for each city. We also include two quantitative positioning variables (latitude and longitude) as well as a categorical variable: Area (categorical variable with the four categories north, south, east, and west of Europe). An extract of the data is provided in Table 1.14.

TABLE 1.14
Temperature Data (Extract): Temperatures Are in Celsius

	Jan	Feb	Mar	Apr	...	Nov	Dec	Ann	Amp	Lat	Lon	Area
Amsterdam	2.9	2.5	5.7	8.2	...	7.0	4.4	9.9	14.6	52.2	4.5	West
Athens	9.1	9.7	11.7	15.4	...	14.6	11.0	17.8	18.3	37.6	23.5	South
Berlin	−0.2	0.1	4.4	8.2	...	4.2	1.2	9.1	18.5	52.3	13.2	West
Brussels	3.3	3.3	6.7	8.9	...	6.7	4.4	10.3	14.4	50.5	4.2	West
Budapest	−1.1	0.8	5.5	11.6	...	5.1	0.7	10.9	23.1	47.3	19.0	East
Copenhagen	−0.4	−0.4	1.3	5.8	...	4.1	1.3	7.8	17.5	55.4	12.3	North
Dublin	4.8	5.0	5.9	7.8	...	6.7	5.4	9.3	10.2	53.2	6.1	North
Helsinki	−5.8	−6.2	−2.7	3.1	...	0.1	−2.3	4.8	23.4	60.1	25.0	North
Kiev	−5.9	−5.0	−0.3	7.4	...	1.2	−3.6	7.1	25.3	50.3	30.3	East
⋮	⋮	⋮	⋮	⋮	⋮	⋮	⋮	⋮	⋮	⋮	⋮	⋮

1.10.2   Analysis Parameters

1.10.2.1   Choice of Active Elements

Cities. We wish to understand the variability of monthly temperatures from one country to another in a multidimensional manner, that is, by taking into account the 12 months of the year simultaneously. Each country will be represented by the climate of its capital. The data of the other cities are not taken into account to avoid giving more weight to the countries for which several cities are listed. Thus, the capitals will be regarded as active individuals while the other cities will be regarded as supplementary individuals (i.e., individuals which are not involved in the construction of the components). From a multidimensional point of view, two cities are all the more close when they present a similar set of monthly temperatures. This data can be summarised by highlighting the capitals' principal components. This will provide answers to questions such as: which are the greatest disparities between countries? These components could be used as a basis for constructing a typology of the countries.

Variables. Each variable measures the monthly temperatures in the 23 capitals: the relationships between variables are considered from the point of view of the capitals alone (i.e., from the active individuals), rather than from all of the cities. The relationships between the variables are a key objective in such a study. Two variables are positively correlated if, overall, the warmest cities according to one variable (the temperature for a given month) are also the warmest according to the other (for example, is it hot in August where it is hot in January?). Naturally, we wish to get a picture of these links, without having to review every pair of variables.

This picture can be obtained using synthetic variables. The question is, can we summarise monthly precipitation with only a small number of components? If so, we will examine the links between the initial variables and the synthetic variables: this indirect review is more convenient than the direct one (with 12 initial variables and 2 synthetic variables, we will examine 24 links instead of $12 \times 11/2 = 66$).

Looking at cities' temperature profiles, we will consequently consider only those variables relating to temperature as active (thus eliminating variables such as latitude and longitude). For the other variables considered supplementary (average annual temperature and annual amplitude), these are synthetic indicators that it will be interesting to compare with the principal components, but which are not part of the profile itself. Moreover, these variables use information already present in the other variables.

1.10.2.2   Should the Variables Be Standardised?

Centring and reducing to unit variance (in other words standardising) is essential only when the active variables are not measured in the same units. The illustrative variables will be analysed through their correlation coefficient with the components and thus automatically standardised.

Coherence of units of measurement does not imply that one does not need to standardise: does 1 degree Celsius represent the same thing in January as in July? Failing to standardise the data means giving each variable a weight proportional to its variance. It must be noted that the standard deviations differ sufficiently little from one month to another (at most they are doubled). It is therefore reasonable to think that standardising has no real impact on the results of the analysis. From another point of view, when calculating distances between towns, failing to standardise the data means attributing the same influence to a difference of 1 degree, whatever the month. When standardised, this difference is amplified as it becomes more evident during the months where the temperatures vary little from one town to another. For this example, it was decided to standardise the data, thus attributing equal weights to each month.

1.10.3   Implementation of the Analysis

Here are the lines of code for obtaining the graphs and the outputs of the analysis that we are commenting:

```
> library(FactoMineR)
> temperature <- read.table("http://factominer.free.fr/bookV2/temperature.csv",
    header=TRUE,sep=";",dec=".",row.names=1)
> res<-PCA(temperature,ind.sup=24:35,quanti.sup=13:16,quali.sup=17)
> plot.PCA(res,choix="ind",habillage=17)
> summary(res)
> dimdesc(res)
> scale(temperature[1:23,1:16])*sqrt(22/23)
> cor(temperature[1:23,1:16])
```

These lines of code allow us:

- to import the data file (specifying that the name of the variables is shown, the field separator is ";", the decimal marker is "." and the name of the individuals is presented in the first column);

- to compute the PCA with the supplementary individuals from 24 to 35 (towns which are not capital cities), the supplementary quantitative variables from 13 to 16, and variable 17 which is categorical supplementary;

- to build the graph of individuals by colouring individuals according to the variable *Area*;

- to retrieve the main results tables: on the variances explained by each component, on active individuals, on illustrative individuals, on (quantitative) active variables, on supplementary quantitative variables, and on the additional categorical variables;

- to describe the dimensions or principal components based on the variables;

- to compute the standardised data for the quantitative variables of the active individuals only; and

- to compute the correlation matrix.

```
> summary(res,nb.dec=2)

Call:
PCA(X = temperature, ind.sup = 24:35, quanti.sup = 13:16, quali.sup = 17)

Eigenvalues
                     Dim.1  Dim.2  Dim.3  Dim.4  Dim.5  Dim.6  Dim.7
Variance              9.95   1.85   0.13   0.04   0.02   0.01   0.01
% of var.            82.90  15.40   1.05   0.32   0.14   0.11   0.05
Cumulative % of var. 82.90  98.29  99.35  99.67  99.81  99.91  99.96
                     Dim.8  Dim.9 Dim.10 Dim.11 Dim.12
Variance              0.00   0.00   0.00   0.00   0.00
```

```
% of var.                 0.02   0.01   0.01   0.00   0.00
Cumulative % of var.     99.98  99.99  99.99 100.00 100.00
```

Individuals (the 10 first)

	Dist	Dim.1	ctr	cos2	Dim.2	ctr	cos2	Dim.3	ctr	cos2
Amsterdam	1.44	0.23	0.02	0.02	-1.37	4.43	0.90	-0.10	0.38	0.01
Athens	7.68	7.60	25.25	0.98	0.93	2.04	0.01	0.56	10.85	0.01
Berlin	0.50	-0.29	0.04	0.33	0.02	0.00	0.00	-0.29	2.91	0.33
Brussels	1.36	0.63	0.17	0.22	-1.18	3.26	0.75	-0.15	0.80	0.01
Budapest	2.45	1.67	1.22	0.46	1.71	6.90	0.49	-0.50	8.57	0.04
Copenhagen	1.63	-1.46	0.93	0.81	-0.49	0.57	0.09	0.44	6.68	0.07
Dublin	2.74	-0.51	0.11	0.03	-2.67	16.82	0.96	-0.18	1.10	0.00
Elsinki	4.13	-4.04	7.12	0.96	0.46	0.50	0.01	0.59	12.12	0.02
Kiev	2.65	-1.71	1.28	0.42	2.01	9.48	0.57	-0.17	1.00	0.00
Krakow	1.57	-1.26	0.69	0.65	0.87	1.80	0.31	-0.27	2.58	0.03

Supplementary individuals (the 10 first)

	Dist	Dim.1	cos2	Dim.2	cos2	Dim.3	cos2
Antwerp	1.33	0.59	0.20	-1.16	0.76	-0.13	0.01
Barcelona	6.01	5.99	0.99	-0.38	0.00	0.25	0.00
Bordeaux	2.97	2.85	0.92	-0.73	0.06	-0.27	0.01
Edinburgh	2.69	-1.29	0.23	-2.34	0.76	-0.18	0.00
Frankfurt	0.68	0.39	0.33	0.15	0.05	-0.44	0.43
Geneva	0.54	0.32	0.36	0.17	0.10	-0.26	0.22
Genoa	5.87	5.85	0.99	-0.13	0.00	0.34	0.00
Milan	3.55	3.16	0.79	1.56	0.19	-0.25	0.00
Palermo	7.41	7.29	0.97	-0.21	0.00	-0.59	0.01
Seville	7.87	7.86	1.00	0.26	0.00	0.22	0.00

Variables (the 10 first)

	Dim.1	ctr	cos2	Dim.2	ctr	cos2	Dim.3	ctr	cos2
January	0.84	7.13	0.71	-0.53	15.28	0.28	0.07	3.64	0.00
February	0.88	7.86	0.78	-0.46	11.25	0.21	0.00	0.01	0.00
March	0.95	8.98	0.89	-0.29	4.47	0.08	-0.12	11.56	0.01
April	0.97	9.53	0.95	0.10	0.54	0.01	-0.20	31.13	0.04
May	0.87	7.61	0.76	0.46	11.34	0.21	-0.16	19.30	0.02
June	0.83	6.98	0.69	0.55	16.09	0.30	0.05	1.94	0.00
July	0.84	7.16	0.71	0.51	14.00	0.26	0.15	18.71	0.02
August	0.91	8.31	0.83	0.40	8.74	0.16	0.09	6.06	0.01
September	0.99	9.77	0.97	0.15	1.26	0.02	0.02	0.41	0.00
October	0.99	9.88	0.98	-0.08	0.39	0.01	0.00	0.00	0.00

Supplementary continuous variables

	Dim.1	cos2	Dim.2	cos2	Dim.3	cos2
Annual	1.00	1.00	-0.07	0.00	0.00	0.00
Amplitude	-0.31	0.10	0.94	0.89	0.04	0.00
Latitude	-0.91	0.83	-0.22	0.05	0.18	0.03
Longitude	-0.36	0.13	0.64	0.42	-0.04	0.00

Supplementary categories

	Dist	Dim.1	cos2	v.test	Dim.2	cos2	v.test	Dim.3	cos2	v.test
East	1.78	-1.10	0.38	-1.08	1.38	0.60	3.15	-0.26	0.02	-2.25
North	2.65	-2.44	0.85	-2.40	-0.99	0.14	-2.26	0.27	0.01	2.35
South	4.57	4.56	1.00	3.58	0.14	0.00	0.25	0.12	0.00	0.80
West	1.03	0.50	0.23	0.34	-0.86	0.69	-1.36	-0.16	0.03	-1.00

The first principal component is predominant since it alone summarises 82.9% of the total inertia (see graphical outputs or the results in `res$eig`). The second principal component is relatively important since it summarises 15.4% of the total inertia. These two principal components express 82.9 + 15.4 = 98.3% of the total inertia which justifies limiting ourselves to the first two components. In other words, from two synthetic variables, we are able to summarise most of the information provided by the 12 initial variables. In this case study, the summary provided by PCA is almost complete. This means that the variables and individuals in the main plane are effectively projected, and that the proximity of two individuals within the plane reflects proximity in the overall space. Similarly, the angle between two variables in the plane gives a very good approximation of the angle in the overall space.

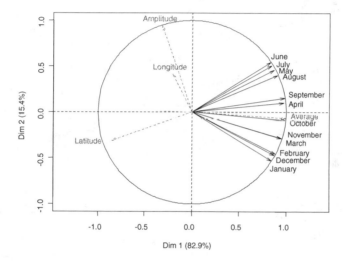

FIGURE 1.20
Temperature data: graph of variables.

First Component
All the active variables are on the same side of the first component (the sign of the correlation coefficient with the first component is the same for all 12 variables, see Figure 1.20). We are dealing with a size effect: some cities have high temperatures regardless of the months of the year, others have low temperatures regardless of the month. In other words, the months are, in general, positively correlated two by two. This first component can be summarised by the term "average annual temperature." This summary is reinforced by a correlation coefficient of 0.998 between this principal component and the illustrative variable average annual temperature (the graph seems to show a weaker correlation but the result in `res$quanti.sup$coord` indicates a correlation of 0.998). Furthermore, we note that the months September, October, and April are more closely linked than the others to this first component: they

"represent" the best the annual temperatures. Apart from the average annual temperature mentioned above, another supplementary quantitative variable is linked to the first principal component: latitude. The correlation between latitude and the first principal component is worth -0.85, which means that the cities that are further south (lower latitude) have a higher coordinate on the first component and are therefore the warmest cities: this is obviously not a surprise!

Remark

The size effect is more informative than the summary *annual temperatures* since it indicates that the hottest cities annually are also (generally) the hottest each month.

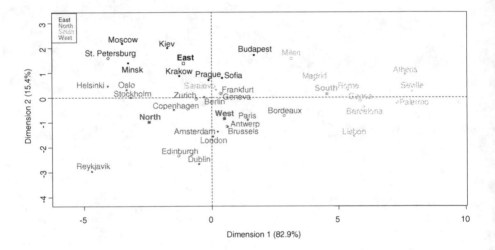

FIGURE 1.21
Temperature data: graph of individuals.

Due to duality, the coordinate of *Helsinki* (*Athens*) reflects a city where it is cold (or hot, respectively) throughout the year (see Figure 1.21). This is clearly visible in the data: whatever the month, *Helsinki* (*Athens*) is a colder than average city (or indeed warmer, respectively). This is more easily visible when working from the standardised data:

```
> scale(temperature[1:23,1:12])*sqrt(22/23)
```

Note that the data is standardised based on active individuals only, that is, from the first 23 individuals.

Second Component

The second component opposes, on the one hand, the months of the period May–July and, on the other hand, those of the period November–March. This can be summed up by the opposition "summer" — "bad season" (i.e., all of the months which cannot be qualified as being part of "summer"). It is important to note that this opposition has nothing to do with an evolution of the averages, since the data are centered prior to analysis. This opposition reflects the fact that, for a given annual temperature, some cities are relatively rather hot in summer and others rather cold. In this case, the word "relatively" is required when interpreting due to the fact that data are centered. The opposition between the cities mentioned above can be found directly in the standardised data; large variations of averages between months may make it difficult to illustrate in the raw data.

Annual thermal amplitude is related to this second component, which can be connected to the two following facts: first, the highest values of this variable were observed for most continental cities (at the top of the axis), and second, the lowest values were observed for the cities nearest to the Atlantic (at the bottom of the axis). *Longitude* is linked to this component, but the relationship is not particularly strong (correlation $= 0.4196$).

Thus, through duality, cities such as *Kiev, Moscow*, or *Budapest* have rather high standardised data during the summer, which is in turn rather low during the rest of the year. On the other hand, cities such as *Dublin* or *Reykjavic* have rather low standardised data values during the summer which are in turn mild during winter. In fact, this opposition may also be read directly in the raw data. This component clearly distinguishes coastal cities with low thermal amplitude from continental cities with high thermal amplitude. The temperatures of the coastal cities (for instance, *Dublin* or *Reykjavic*) are generally medium or low (indicated by the first component) and are very low during the summer. In contrast, cities located inland (such as *Kiev, Moscow*, or *Budapest*) are generally medium or low and very low during winter and high during summer.

The automatic description of the components obtained from quantitative variables (using the **dimdesc** function) confirms our interpretation of the components. The categorical variable *Area* also makes it possible to characterise the components. The categories *North* and *South* characterise the first component: the category *South* (*North*) has a significantly positive coordinate (negative) on the first component. This is interpreted as meaning that Southern European cities (Northern) are warmer (or colder, respectively) throughout the year. The category *East* (*North*) has a significantly positive coordinate (negative) on the second component. This can be interpreted as: Eastern European cities (Northern) have high (or, respectively, low) thermal amplitudes.

```
> dimdesc(res)
$Dim.1
$Dim.1$quanti
          correlation   P-value
Average         0.998   9.58e-26
October         0.992   3.73e-20
September       0.986   1.06e-17
April           0.974   5.30e-15
```

```
November        0.952   2.66e-12
March           0.945   1.15e-11
August          0.909   1.90e-09
February        0.884   2.18e-08
December        0.873   5.45e-08
May             0.870   7.01e-08
July            0.844   4.13e-07
January         0.842   4.59e-07
June            0.833   7.96e-07
Latitude       -0.852   2.57e-07


$Dim.1$quali
             R2      P-value
Area       0.679   6.282e-05

$Dim.1$category
         Estimate    P-value
South      4.183    2.28e-05
East      -1.478    4.09e-02
North     -2.823    4.98e-04

$Dim.2
$Dim.2$quanti
           correlation   P-value
Amplitude      0.944   1.30e-11
June           0.545   7.11e-03
July           0.509   1.32e-02
May            0.458   2.80e-02
Longitude      0.420   4.62e-02
February      -0.456   2.88e-02
December      -0.473   2.27e-02
January       -0.531   9.08e-03

$Dim.2$quali
         Estimate    P-value
Area       0.546    0.00153

$Dim.2$category
         Estimate    P-value
East       1.462    4.47e-04
North     -0.906    1.66e-02
```

Conclusion

Assessing the relationships between temperatures revealed positive correlations between monthly temperatures and, more precisely, two periods: the summer season (May to August) and the bad season (November to March). Temperatures are more closely linked within each period than from one period to the other. Temperatures can be summarised by two synthetic variables: average annual temperature and thermal amplitude. From these two variables, we can outline the city typologies. By bringing together those cities which are close on the map defined by the first two components, and respecting the geographical location, we can propose the following typology:

- Southern European cities are characterised by high temperatures throughout the year.

- Western European cities are characterised by average temperatures throughout the year.

- Northern European cities are characterised by cold temperatures, especially during summer.

- Eastern European cities are characterised by cold temperatures, especially during winter.

The temperature profile for the Southern European town of *Sarajevo* is more similar to that of towns in Western Europe than it is to those in Southern Europe.

It may be noted that cities that have not participated in the construction of the components (the supplementary individuals in the analysis) have similar temperature profiles to the capitals of the countries they belong to.

The variables *November* and *March* are strongly correlated: indeed, the ends of the arrows are close to the circle of correlation, so the angle between vectors *November* and *March* in the space $\mathbb{R}^K$ (space of the individuals) is close to the angle on the plane, namely, close to 0. As the correlation coefficient is the cosine of the angle in the individuals' space, the correlation coefficient is close to 1. This means that the cities where it is cold in November are also those where it is cold in March.

The correlation between *January* and *June* is close to 0 because, on the plane, the angle is close to $\pi/2$ and the variables are well projected.

1.11   Example of Genomic Data: The Chicken Dataset

1.11.1   Data Description — Issues

The data, kindly provided by S. Lagarrigue, relates to 43 chickens having undergone one of the six following diet conditions: normal diet (N), fasting for 16 hours (F16), fasting for 16 hours then refed for 5 hours (F16R5), fasting for 16 hours then refed for 16 hours (F16R16), fasting for 48 hours (F48), and fasting for 48 hours then refed for 24 hours (F48R24). At the end of the diet, the genes were analysed using DNA chips, and the expression of 7407 genes retained for all the chickens. A biologist selected the most pertinent genes, since at the beginning, more than 20,000 genes were identified by DNA chips. The data were then preprocessed in a standard manner for DNA chips (normalisation, eliminating the chip effect, etc.).

The data table to be analysed is a rectangular array with far fewer individuals than variables: 43 rows (chicken) and 7407 columns (genes). In addition, there is also a categorical variable, *Diet*, which corresponds to one of the six stress situations, or diets, outlined above.

The aim of the study is to see whether the genes are expressed differently depending on the situation of stress (a situation of stress = a mode) which the chicken is subject to. More precisely, it may be interesting to see how long the chicken needs to be refed after fasting before it returns to a normal state, i.e., a state comparable to the state of a chicken with a normal diet. Might some genes be underexpressed during fasting and overexpressed during feeding?

1.11.2   Analysis Parameters

Choice of active elements: in this study, all the chickens are regarded as active individuals and all genes as active variables. The variable *Diet* is illustrative since it is categorical.

Should the variables be standardised? In our case, variables are standardised to attribute the same influence to each gene.

1.11.3   Implementation of the Analysis

Technical difficulties can arise when importing this type of data file with many columns. For example, it is impossible to import the dataset in OpenOffice Calc (version 3.0) since it does not support more than 1024 columns. It is therefore common for the table to be formatted with genes as rows and individuals as columns. The categorical variable *Diet* should not be entered into this table, otherwise, during importation, all the variables would be regarded as categorical (for a variable, if one piece of data is categorical, the entire variable is regarded as categorical). The data table (gene × chicken) can be imported and then transposed by running the following code:

```
> chicken <- read.table("http://factominer.free.fr/bookV2/chicken.csv",header=TRUE,
     sep=";",dec=".",row.names=1)
> chicken <- as.data.frame(t(chicken))
```

The categorical variable *Diet* must then be merged with the data table (once it is created):

```
> diet <- as.factor(c(rep("N",6),rep("F16",5),rep("F16R5",8),rep("F16R16",9),
     rep("F48",6),rep("F48R24",9)))
> chicken <- cbind.data.frame(diet,chicken)
> colnames(chicken)[1] <- "Diet"
```

We can then carry out the PCA, summarize the main results, and build the graph of individuals by colouring the individuals according to the variable *Diet* (here the first variable in the table). The size of the text can be changed using the setting `cex` ("cex = 0.7" instead of 1 by default):

```
> res.pca <- PCA(chicken,quali.sup=1)
> summary(res.pca)
> plot(res.pca,habillage=1,cex=0.7)
```

The principal plane expresses 29.1% of the total inertia (see the graphs or the object **res.pca$eig**). Note that here we obtain 42 dimensions at most, which corresponds to the number of individuals −1 (and not the total number of variables): the 43 individuals are therefore in a space with 42 dimensions at most.

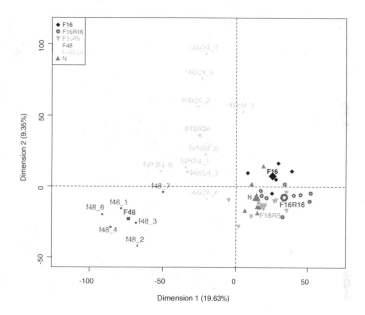

FIGURE 1.22
Genomic chicken data: graph of individuals on the first plane.

The principal plane of the PCA (see Figure 1.22) separates the chickens into two subgroups. Those that have undergone intense stress (48 hours of fasting), are greatly dispersed, and those that have been subject to less intense stress (16 hours of fasting), are more concentrated and located close to the origin. Furthermore, the first component separates the chickens into three groups: chickens that have undergone intense stress but that have not been refed afterwards (F48), chickens that have undergone intense stress and that have been refed afterwards (F48R24), and the other chickens. Chickens which have been refed tend to recover from intense stress and their health tends to be similar to that of normal chickens. However, 24 hours of feeding is not enough for the state of the chicken to completely return to normal. This means that some genes are specific to a state of intense stress, as some genes are overexpressed under stress while others are underestimated (the graph of variables shows that certain variables are negatively correlated while others are positively correlated). The second component is specific to the chickens F48R24.

The graph of variables is unclear here due to the large number of variables (see Figure 1.23). To represent this graph, and to examine whether or not there is a particular structure to the variables, we provide one point per variable (with no arrows or labels) using the command

```
> plot(res.pca,choix="var",invisible="var")
> points(res.pca$var$coord[,1:2], cex=0.5)
```

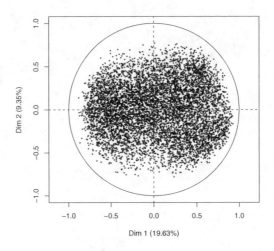

FIGURE 1.23
Genomic chicken data: graph of individuals on the main plane of variability (each point corresponds to a variable).

This cloud is regularly distributed and does not need to be commented (but nonetheless had to be verified). It is therefore necessary to characterise the components using the **dimdesc** function (here, we provide only the variables that are the most characteristic of quantitative variables, and the set of categories that are the most characteristic of quantitative variables):

```
> dimdesc(res.pca,proba=1e-5)
```

$Dim.1$quanti		$Dim.2$quanti		$Dim.3$quanti	
	Dim.1		Dim.2		Dim.3
HS2ST1	0.93	MPHOSPH9	0.77	⋮	⋮
TTC151	0.92	BNC2	0.76	AGL	−0.83
PRC1	0.91	XRCC6	0.75	LRRC8A	−0.83
KLHL101	0.91	FBXW4	0.75	ARFIP1	−0.84
C6orf66	0.91	OFD11	0.75	PRDM161	−0.85
C16orf48	0.91	USP53	0.73	PDE4B	−0.86
TTC8	0.91	⋮	⋮	GLI2	−0.87
KCNJ15	0.90	DNAH10	−0.75	PRKAA2	−0.87
GRIA3	0.90	RHOT2	−0.76	PCSK51	−0.89
C7orf30	0.90	PDCD11	−0.77	DUSP14	−0.89

```
   ⋮         ⋮      PHYHD1   -0.78     HIPK2    -0.90

$Dim.1$quali        $Dim.2$quali        $Dim.3$quali
        Dim.1               Dim.2               Dim.3
F16R16    2.98     F48R24   4.59     F16       3.58
F48R24   -2.24     F48     -2.25     F16R5     2.30
F48      -5.02                       N        -3.85
```

For the first component, the genes that are the most correlated to that component are all positively correlated. Thus, these genes are underexpressed when the chickens have been fasting for 48 hours. For component 2, some genes are overexpressed (MPHOSPH9, BNC2, etc.) when the chickens are refed after fasting for 48 hours while other genes are underexpressed (PHYHD1, PDCD11, etc.). Of course, here the statistician needs to ask a biologist to analyse why these particular genes are under- or overexpressed. Several categories of the variable *Diet* are characteristic of components 1 and 2: here we find a result that we see on the plane, but only by using an additional test (while visually we cannot say whether the differences are significant or not). The chickens that suffered 48 hours of stress have significantly lower coordinates than the others on component 1, whether refed afterwards or not. Inversely, chickens that have suffered 16 hours of stress and been refed afterwards have significantly positive coordinates. Component 2 is characterised by chickens subjected to 48 hours of stress and opposes chickens refed afterwards (with significantly positive coordinates) with those that were not (with significantly negative coordinates).

It is also possible to visualise plane 3-4 of the PCA:

```
> plot(res.pca,habillage=1,axes=3:4)
> plot(res.pca,choix="var",invisible="var",axes=3:4)
> points(res.pca$var$coord[,3:4],cex=0.5)
```

Plane 3-4 (see Figure 1.24), and more specifically component 3, allows us to differentiate the diets that have not been differentiated on the principal plane. Chickens that followed a normal diet have negative coordinates on component 3 and chickens who suffered 16 days of stress have positive coordinates on the same component. Chickens that were refed after 16 days of stress are between these two groups, with a gradient depending on duration of feeding time: chickens refed for 5 hours are closer to those that were not refed, and chickens refed for 16 hours are closer to those that did not suffer stress. It therefore seems that some genes are expressed differently according to whether there was a stress of 16 hours or not, when some genes return gradually to a "normal" state. However, even after 16 hours of feeding, the genes do not function normally again.

As for the principal plane, the cloud of variables on plane 3-4 is regularly distributed and does not need to be commented (see Figure 1.25). It is easier to characterise the components automatically using the procedure **dimdesc**. Component 3 is characterised by variables HIPK2, DUSP14, or

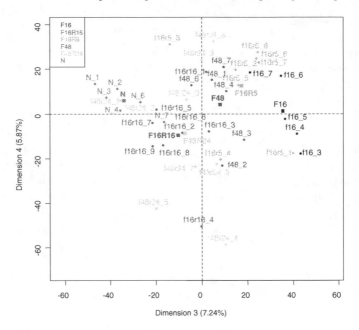

FIGURE 1.24
Genomic chicken data: graph of individuals on plane 3-4.

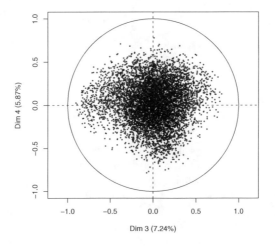

FIGURE 1.25
Genomic chicken data: graph of individuals on the plane stemming from components 3-4 (each point corresponds to a variable).

PCSK51: these are the genes which are the most closely related to the component (negative correlation). The categories that characterise the component are associated with chickens that were not subject to stress (with a significantly negative coordinate), chickens which underwent 16 hours of stress and were not refed afterwards and, to a lesser extent, chickens that suffered 16 hours of stress and that were refed afterward for 5 hours (with a significantly positive coordinate).

It is possible to draw confidence ellipses around the barycentre of the representation of all the chickens having undergone the same diet. To do so, we use the function **plotellipses** with the argument `bary=TRUE`, which specifies that the ellipses are constructed around the barycentres. This argument is used by default and does not need to be specified:

```
> plotellipses(res.pca)
```

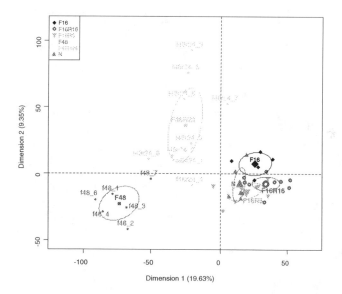

FIGURE 1.26
Genomic chicken data: confidence ellipses around the categories of the variable *Diet* on the main plane.

These confidence ellipses (see Figure 1.26) confirm the visual impression that the chickens under excessive stress (F48 and F48R24) are very different from the others. However, one can also see that the confidence ellipses are disjointed for diets F16 and F16R16, F16R16 and N, or F16 and F16R5: this differentiation of the diets was not at all obvious without the confidence ellipses. To obtain the confidence ellipses on components 3-4, we use

```
> plotellipses(res.pca, axes=3:4)
```

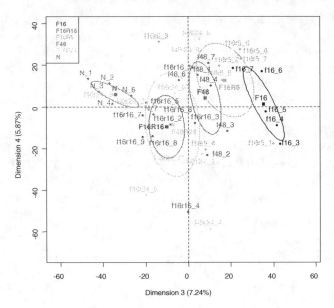

FIGURE 1.27
Genomic chicken data: confidence ellipses around the categories of the variable
Diet on the plane stemming from components 3-4.

On plane 3-4, several categories of the variable *Diet* are clearly distinguishable (see Figure 1.27): diet N is different from all the others and in particular
from F16R16. This indicates that the chickens which underwent 16 days of
stress and which were refed for the following 16 days still did not recover from
their stress.

2

Correspondence Analysis (CA)

2.1   Data — Notation — Examples

To illustrate this chapter, we will be working from a small data table featuring an extract of the results of a survey which, though not recent, is rather remarkable.[1] Completed by 1724 women, a long questionnaire, which, among other things, featured a battery of questions relating to their attitude toward women's work. This data represents a turning point in history in sociological terms. The end of the 1960s also marked the end of a number of feminist struggles, particularly regarding women's access to paid work (in France, women could not work without their husband's consent before 1965). It is for this reason that a number of studies concerning women's opinions were conducted at this time. Our illustration features two questions; the headings for these questions and the responses can be found in Table 2.1.

TABLE 2.1
Table Showing the Responses to Two Opinion Questions

In an ideal family:	The most suitable activity for a mother when the children go to school			
	Stay at home	Part-time work	Full-time work	
Both parents work equally	13	142	106	261
Husband works more	30	408	117	555
Only the husband works	241	573	94	908
Total	284	1123	317	1724

This table is known as a "cross tabulation" used by organisations conducting surveys, and as a "contingency table" by statisticians. It summarises the responses to both questions: 241 therefore corresponds to the number of people who replied *only the husband works* to the question about an ideal family, and *stay at home* to the question about the best activity for a mother. The table is supplemented by the sum of terms on any one row (908 people replied *only the husband works*; these numbers are therefore the column margin) or any one column (284 people chose the answer *stay at home*; these numbers therefore correspond to the row margin). The questions were worded as follows:

---

[1]Tabard N. (1974). *The needs and hopes of families and young people*, CREDOC, Paris.

Out of the following three options, which best corresponds to your image of the ideal family:

1. A family where both parents each have a profession which interests them equally and where housework and childcare are equally shared.

2. A family where the mother has a profession, but which is less time-consuming than that of the father, and where she carries out the majority of the housework and childcare.

3. A family where only the husband works and where the wife stays at home.

When identifying a difference between the period when children are very young and when they start school, what do you think is the best activity for a mother:

1. Stay at home

2. Part-time work

3. Full-time work

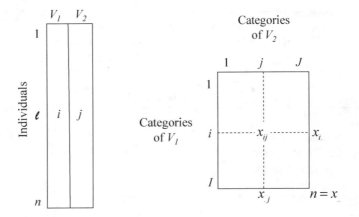

FIGURE 2.1
General indications for a contingency table with the two categorical variables V_1 and V_2 defined for n individuals; individual l carries the categories i (for V_1) and j (for V_2): these are accounted for in x_{ij}.

More generally, a contingency table is constructed as follows (see Figure 2.1). For n individuals, the values are available for two categorical variables: V_1 (with I categories or levels) and V_2 (with J categories). The contingency table carries the general term x_{ij} as the number of individuals featuring in the category i of V_1 and j of V_2.

The sums of the table are referred to as the margin and are denoted by replacing the index in x_{ij} upon which the calculation is made by a point. Thus

$$x_{i\bullet} = \sum_{j=1}^{J} x_{ij} \quad x_{\bullet j} = \sum_{i=1}^{I} x_{ij} \quad n = x_{\bullet\bullet} = \sum_{i,j} x_{ij}.$$

Finally, in correspondence analysis, we consider probability tables[2] associated with contingency tables as the general term $f_{ij} = x_{ij}/n$, the probability of carrying both the categories i (of V_1) and those of j (V_2). The margins of this table, which are also known as marginal probabilities, are defined by

$$f_{i\bullet} = \sum_{j=1}^{J} f_{ij} \quad f_{\bullet j} = \sum_{i=1}^{I} f_{ij} \quad f_{\bullet\bullet} = \sum_{i,j} f_{ij} = 1.$$

Remark

The term "correspondence analysis" (CA) stems from the fact that tables are analysed by linking two corresponding sets: that which is represented by the rows, and that represented by the columns (they have symmetrical roles).

2.2 Objectives and the Independence Model

2.2.1 Objectives

Survey results are traditionally summarised by counting the responses to a series of (carefully chosen) questions. In the example, the answers to question 1 (on the ideal family) clearly show that women (in France in 1970, a detail that we cannot repeat enough) were generally against the idea of women working (52.7% chose *only the husband works*). However, the responses to question 2 (about the ideal activity for mothers) show, equally clearly, that women are largely in favour of the idea of women working (only 16.47% women chose the response *stay at home*). We might therefore conclude that simply counting the number of responses to only one question in an opinion poll can yield only flimsy results (which some say are of little or no value). We must therefore simultaneously account for the answers to several questions (two in this chapter, and more than two in the following chapter, which deals with multiple correspondence analysis). In our example, we hope that the cross tabulation of the answers to both questions will help us to understand the contradictions which arise when considering each question separately.

Empirically speaking, analysing this table consists of comparing the figures. If it were simply a series of nine numbers, we would focus on the greatest

---

[2]In this example, the term "probability" may seem inappropriate as it refers to an amount determined from a sample. However, as well as being convenient, it also corresponds well to the fact that, in correspondence analysis, data are considered as populations inasmuch as they are considered without inferences.

and the smallest values. Thus, the greatest value in the table, 573, seems to suggest an "attraction" between the categories *only the husband works* and *part-time work*. It appears to be confirmed by the fact that *part-time work* is the most common answer for those people who replied *only the husband works* and vice versa. However, when we consult the margins, it can be seen that these two answers, when considered separately, both represent the majority within their set of categories. This raises the following question: is the value of 573 only high because these responses each occur frequently, rather than due to some "attraction" between the categories? Here it would appear that the figures in a contingency table can only be compared to one another by accounting for their corresponding margins. Analysing a table such as this is thus rather tricky: we must therefore establish the objective of the study jointly with a methodology in use.

2.2.2   Independence Model and χ^2 Test

The main principle behind the construction of a contingency table implies that the aim of studying a table such as this is to examine the relationships between the answers to two questions. It must be noted here that, as in most tables used in CA, we are sure that a relationship exists. In light of the significance of these questions, conducting a χ^2 test and finding that there is no relationship between the questions in Table 2.1 would be something of a shock, and would call the quality of the data into question.

Studying the link between two variables requires us to position the data in terms of a given starting point, in this case the absence of a relationship. The independence model specifies this criterion as a starting point. The standard relationship of independence between two events (P[A and B]=P[A]P[B]) is applicable to both categorical variables. Two categorical variables are considered independent if they verify

$$\forall i, j \quad f_{ij} = f_{i\bullet} f_{\bullet j}.$$

As a result, the independence model stipulates that the joint probability (f_{ij}) is dependent on the marginal probabilities ($f_{i\bullet}$ and $f_{\bullet j}$) alone, which seems to correspond to our remark regarding the value 573.

Studying these relationships means comparing the actual sample sizes ($x_{ij} = n f_{ij}$) and the theoretical sample sizes characterised by the independence model ($n f_{i\bullet} f_{\bullet j}$). Table 2.2 combines these two tables to illustrate our example.

Let us comment on some of the differences between the table of actual samples and that of theoretical samples:

- 13 women replied both *both parents work equally* and *stay at home*: if the questions had been independent, 43 people (on average) would have chosen this pair of answers. The actual sample is very slightly smaller than the theoretical sample, which is to be expected, given the nature of the answers.

TABLE 2.2
Actual Samples and Theoretical Samples

Actual Sample Size

	Stay at home	Part-time work	Full-time work	Total
Both work equally	13	142	106	261
Husband works more	30	408	117	555
Only husband works	241	573	94	908
Total	284	1123	317	1724

Theoretical Sample Size

	Stay at home	Part-time work	Full-time work	Total
Both work equally	43.0	170.0	48.0	261
Husband works more	91.4	361.5	102.1	555
Only husband works	149.6	591.5	167.0	908
Total	284	1123	317	1724

We say that these categories repel one another: when subjects choose one of the answers, they are unlikely to choose the other.

- 241 people chose both *only the husband works* and *stay at home*, a value which is considerably higher than the (average) theoretical value of 149.6 obtained using the independence hypothesis (again, this result is expected, given the nature of the answers). We say that these categories attract one another: when a subject chooses one of the answers, he is likely to choose the other.

- 573 people chose both *only the husband works* and *part-time work*, a sample which is (very slightly) lower than the theoretical sample of 591.5.

The above results are interesting from a methodological perspective. The highest value within the table, 573, suggests to the untrained eye a relationship between the two answers. However, this is not the case because, quite the contrary, these categories repel one another (although weakly). The high value of 573 can therefore be attributed to the fact that both categories (when considered separately) have high frequencies (52.7% and 65.1% of the responses, respectively) rather than to the fact that they are related in any way. This result, which could have been predicted, is here clearly quantified using formalisation (relationship between two variables; divergence from the independence model).

The χ^2 test is used to test the significance of the overall deviation of the actual sample from the independence model. It is expressed as follows:

$$\chi^2 = \sum_{i,j} \frac{(\text{Actual Sample Size} - \text{Theoretical Sample Size})^2}{\text{Theoretical Sample Size}},$$

$$\chi^2 = \sum_{i,j} \frac{(nf_{ij} - nf_{i\bullet}f_{\bullet j})^2}{nf_{i\bullet}f_{\bullet j}} = n \sum_{i,j} \frac{(f_{ij} - f_{i\bullet}f_{\bullet j})^2}{f_{i\bullet}f_{\bullet j}} = n\Phi^2,$$

where $f_{i\bullet}$ can be interpreted as the frequency of individuals who have chosen the category i for variable V_1, $f_{\bullet j}$ can be interpreted as the frequency of individuals who have chosen the category j for variable V_2, and Φ^2 is a measurement of the relationship independent of the sample size and total inertia (see below). In the example, the value of χ^2 is 233.43, which is a highly significant value (p-value: 2.4×10^{-49}). This is to be expected, given the nature of the questions. A more detailed look at the calculations (see Table 2.3) illustrates the contribution of each cell to the deviation from independence (it is the association between *both parents work equally* and *full-time work* which expresses the greatest deviation from independence: 30.04%, of the total) as well as that of the rows and columns (note the very weak contribution of *part-time work*, 4.78%).

TABLE 2.3
χ^2 Decomposition by Cell, Row, and Column (Raw Values and Percentages)

	Stay at home	Part-time work	Full-time work	Total
Both work equally	20.93	4.62	70.12	95.66
Husband works more	41.27	5.98	2.19	49.44
Only husband works	55.88	0.58	31.88	88.34
Total	118.07	11.17	104.19	233.43

	Stay at home	Part-time work	Full-time work	Total
Both work equally	8.96	1.98	−30.04	40.98
Husband works more	17.68	−2.56	−0.94	21.18
Only husband works	−23.94	0.25	13.66	37.84
Total	50.58	4.78	44.63	100.00

Note: For each "signed" value, we attribute the sign when the observed sample size is less than that of the theoretical sample.

2.2.3   The Independence Model and CA

Contingency tables must therefore be analysed in terms of independence. This is done by CA in expressing the independence model as follows:

$$\forall i, j \quad \frac{f_{ij}}{f_{i\bullet}} = f_{\bullet j}.$$

The quantity $f_{ij}/f_{i\bullet}$ is the conditional probability of carrying category j (for variable 2) when carrying category i (for variable 1). Therefore, in all cases, independence arises when the conditional probability is equal to the marginal probability. This perception of independence is similar to what we might expect to find: independence arises if the probability of carrying j (from V_2) does not depend on the category carried by V_1.

In symmetrical terms, the independence model can be expressed as

$$\forall i, j \quad \frac{f_{ij}}{f_{\bullet j}} = f_{i\bullet}.$$

CA simultaneously accounts for both terms, using the terminology from the row profile $\{f_{ij}/f_{i\bullet} \; ; \; j = 1, J\}$, the column profile $\{f_{ij}/f_{\bullet j} \; ; \; i = 1, I\}$, and the average profile (row or column) for the distribution of the entire population for one variable, that is to say $\{f_{i\bullet} \; ; \; i = 1, I\}$ and $\{f_{\bullet j} \; ; \; j = 1, J\}$. The independence model therefore requires that both the row profiles and the column profiles be equal to the corresponding average profile.

2.3 Fitting the Clouds

2.3.1 Clouds of Row Profiles

From the table of row profiles, we build a cloud of points denoted N_I, within the space $\mathbb{R}^J$ in which each dimension corresponds to a category of the variable V_2. This construction is similar to that of the clouds of individuals in Principal Component Analysis (PCA). Each row i has a corresponding point whose coordinates for dimension j is $f_{ij}/f_{i\bullet}$. This cloud is supplemented by the mean point G_I, with a jth coordinate value of $f_{\bullet j}$ (see Figure 2.2).

Besides the transformations into profiles, compared to the cloud of individuals in PCA, the cloud of the rows in CA has the following two characteristics:

1. Each point i is attributed a weight $f_{i\bullet}$; this weight is imposed and represents an integral part of CA. For a given profile, we attribute an influence to each category which increases with its frequency. With this weight, the average profile G_I is the centre of gravity of N_I. This point G_I is taken to be the origin of the axes (as for individuals in PCA).

2. The distance attributed to the space $\mathbb{R}^J$ consists of attributing the weight $1/f_{\bullet j}$ to the dimension j. The square of the distance (called the χ^2 distance) between points i and l is expressed as

$$d_{\chi^2}^2(i, l) = \sum_{j=1}^{J} \frac{1}{f_{\bullet j}} \left(\frac{f_{ij}}{f_{i\bullet}} - \frac{f_{lj}}{f_{l\bullet}} \right)^2.$$

The main justification for this distance is indirect and lies in the following property. When considering the weight $f_{i\bullet}$, the inertia of point i with respect to G_I is expressed as

$$\text{Inertia}(i/G_I) \quad = \quad f_{i\bullet} \, d_{\chi^2}^2(i, G_I) = f_{i\bullet} \sum_{j=1}^{J} \frac{1}{f_{\bullet j}} \left(\frac{f_{ij}}{f_{i\bullet}} - f_{\bullet j} \right)^2,$$

$$\text{Inertia}(i/G_I) \quad = \quad \sum_{j=1}^{J} \frac{(f_{ij} - f_{i\bullet} f_{\bullet j})^2}{f_{i\bullet} f_{\bullet j}}.$$

Up to the multiplicative factor n, we recognise the contribution of the row i to the χ^2 test, which is why it is referred to as the χ^2 distance. The result is that the total inertia of the cloud of points N_I with respect to G_I is equal (up to the multiplicative factor n) to the χ^2 criterion. It may also be said that this inertia is equal to Φ^2. Examining the distribution of N_I in terms of G_I means examining the difference between the data and the independence model. This is what CA does in highlighting the directions of greatest inertia for N_I.

Comment on the Total Inertia of N_I

This inertia, equal to Φ^2, offers us vital information as it measures the intensity of the relationship between the two variables of the contingency table. This differs from standardised PCA in which the total inertia is equal to the number of variables and does not depend on the data itself.

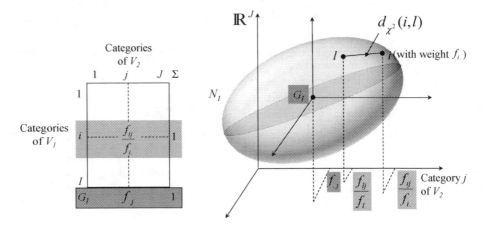

FIGURE 2.2
The cloud of row profiles in CA.

2.3.2   Clouds of Column Profiles

In contingency tables, rows and columns play symmetrical roles: we can study either $V_1 \times V_2$ or $V_2 \times V_1$. This is one of the major differences between CA and PCA in which rows (individuals) and columns (variables) are not analysed in the same manner: we consider distances between individuals and correlations between variables. Consequently, CA constructs the cloud of column profiles in the same way as it builds the cloud of the row profiles (see Figure 2.3):

1. We consider column profiles (thus, depending whether we are interested in the rows or the columns, we do not analyse the same table; $f_{ij}/f_{i\bullet}$ in one case, $f_{ij}/f_{\bullet j}$ in the other. This is one of the

major differences between CA and PCA, in which the same data transformation — centring and reduction — is used for studying both individuals and variables).

2. Each column has a corresponding point in $\mathbb{R}^I$ of which the coordinate on the dimension i is $f_{ij}/f_{\bullet j}$. These points make up the cloud N_J.

3. Each point j is attributed a weight $f_{\bullet j}$. Using this weight, the cloud's centre of gravity, denoted G_J, is equal to the average profile. The origin of the axes is located at G_J.

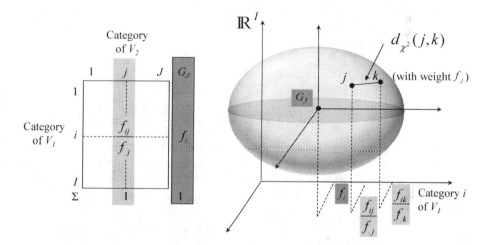

FIGURE 2.3
The cloud of column profiles in CA.

In $\mathbb{R}^I$, the distance attributes the weight $1/f_{i\bullet}$ to the dimension i. The distance (squared) between two columns j and k is expressed as

$$d_{\chi^2}^2(j,k) = \sum_{i=1}^{I} \frac{1}{f_{i\bullet}} \left(\frac{f_{ij}}{f_{\bullet j}} - \frac{f_{ik}}{f_{\bullet k}} \right)^2.$$

Inertia in column j with respect to point G_J is expressed as

$$\text{Inertia}(j/G_J) = f_{\bullet j}\, d_{\chi^2}^2(j,G_J) = f_{\bullet j} \sum_{i=1}^{I} \frac{1}{f_{i\bullet}} \left(\frac{f_{ij}}{f_{\bullet j}} - f_{i\bullet} \right)^2$$

$$= \sum_{j=1}^{J} \frac{(f_{ij} - f_{i\bullet}f_{\bullet j})^2}{f_{i\bullet}f_{\bullet j}}.$$

We can recognise the contribution (up to a multiplicative factor n) of column j to χ^2. The total inertia of N_J is thus the same as that of N_I

$(= \frac{1}{n}\chi^2)$: studying the distribution of N_J around G_J means examining the relationship between both variables V_1 and V_2.

2.3.3   Fitting Clouds N_I and N_J

We work in much the same way as when fitting the cloud of individuals in PCA (see Section 1.3.2). The steps to be followed to construct the cloud of row profiles are detailed below.

The origin of the axes is located at the centre of gravity G_I of cloud N_I evolving within the space $\mathbb{R}^J$. We are aiming for a sequence of orthogonal axes with maximum inertia. Let u_s be the unit vector for the dimension of rank s and H_i^s the projection of the profile i on this dimension. u_s thus maximises the following quantity:

$$\sum_{i=1}^{I} f_{i\bullet} (OH_i^s)^2 \quad \text{maximum.}$$

Cloud N_I is projected on the axes u_s. We represent these projections on the planes which associate two axes (for example, the plane with u_1 and u_2). As in PCA, due to the orthogonality between the axes, this first plane also maximises the projected inertia of N_I: in other words, we obtain the same plane when searching for the plane of maximum inertia directly (rather than axis by axis). Thus, solutions are nested and the best dimension is included in the best plane.

Total inertia measures the intensity of the relationship (in terms of Φ^2) between the two variables V_1 and V_2, and the inertia λ_s associated with the dimension of rank s measures the part of that relationship expressed by that dimension. The inertia λ_s depends on the coordinates of the row profiles on the dimension s: the distance between a profile and the origin can be interpreted as a deviation from the average profile and therefore contributes to the relationship between V_1 and V_2. The proximity of two row profiles i and l also expresses a similar deviation from the average profile. These categories, i and l (of V_1), are preferentially associated (i.e., more than if they were independent) to the same categories of V_2. Similarly, for the same categories of V_2, i and l are less well associated than in the independence model. The fact that both row profiles i and l are opposed with respect to the origin expresses two opposing ways of moving away from the average profile: the categories of V_2 with which i is preferentially associated are also those with which l is less associated, when compared to independence.

CA is conducted in a symmetrical manner to fit cloud N_J. The main stages of this procedure are summarised below. In $\mathbb{R}^I$, the origin of the axes is located at G_J, the centre of gravity of N_J. We are aiming for a sequence of orthogonal axes with maximum inertia. Let v_s be the unit vector for the dimension of rank s and H_j^s the projection of the profile j on this dimension.

v_s thus maximises the following quantity:

$$\sum_{j=1}^{J} f_{\bullet j} \left(OH_j^s\right)^2 \quad \text{maximum.}$$

Cloud N_J is projected on the planes made up of pairs (v_s, v_t), and particularly the first of these pairs (v_1, v_2).

Comment on the Number of Dimensions

As cloud N_I evolves within the space $\mathbb{R}^J$, we may be led to believe that, generally, J dimensions are required for it to be perfectly represented. In reality, two other elements intervene:

1. The sum of the coordinates for a profile is equal to 1; cloud N_I therefore belongs to a subspace of dimension $J - 1$.

2. Cloud N_I is made up of I points; it is possible to represent all these points with $I - 1$ dimensions.

Finally, the maximum number of dimensions required to perfectly represent N_I is therefore $\min\{(I - 1), (J - 1)\}$. The same value is obtained when we work from N_J.

Comment on the Implementation of the Calculations

It can be shown that a matrix diagonalisation is at the heart of CA, and its eigenvalues represent the projected inertia. This is why the term "eigenvalue" appears in the listing in place of "projected inertia": as they represent inertias, these eigenvalues are positive (we will see that they are less than 1) and are organised in descending order (the first dimension corresponds to the maximum projected inertia). The coordinates of the rows and columns are inferred from the eigenvectors associated with these eigenvalues. The rank of this matrix is a priori $\min\{I, J\}$: calculation time therefore mainly depends on the smallest dimension of the table being analysed (as is the case in PCA). In fact, the rank is smaller than $\min\{I - 1, J - 1\}$ due to the χ^2 distance being centred, and hence a linear dependence being introduced.

2.3.4 Example: Women's Attitudes to Women's Work in France in 1970

The CA applied to Table 2.1 yields the two graphs in Figure 2.4. Given the size of the table (3×3), one plane suffices to perfectly represent both clouds. We shall limit our interpretation to the first dimension. The commentary may begin either with the rows or with the columns. We shall illustrate our interpretation of the CA using the tables of row and column profiles (see Table 2.4).

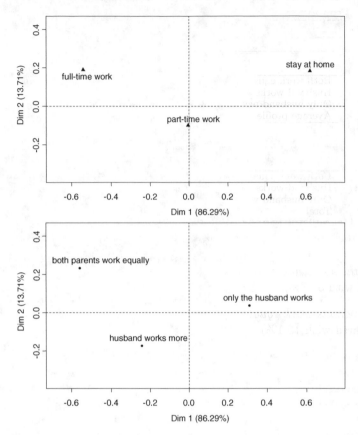

FIGURE 2.4
Planes defined by the first (and the only) two dimensions resulting from a CA of Table 2.1. Above: representation of the columns; below: representation of the rows.

2.3.4.1   Column Representation (Mother's Activity)

The first dimension opposes the categories *stay at home* and *full-time work*. This opposition on the graph inevitably represents an opposition in terms of profile: thus, women who answered *stay at home* (profile column 1) responded:

- *Only the husband works* more often than the average population (= average column profile): 84.9% compared with 52.7%.

- *Both parents work equally* less often than the average population (4.6% compared with 15.1%).

Similarly, women who answered *full-time work* responded:

TABLE 2.4
Row Profiles and Column Profiles for Table 2.1

Row Profiles

	Stay at home	Part-time work	Full-time work	Total
Both work equally	0.050	0.544	0.406	1.000
Husband works more	0.054	0.735	0.211	1.000
Only husband works	0.265	0.631	0.104	1.000
Average profile	0.165	0.651	0.184	1.000

Column Profiles

	Stay at home	Part-time work	Full-time work	Total
Both work equally	0.046	0.126	0.334	0.151
Husband works more	0.106	0.363	0.369	0.322
Only husband works	0.849	0.510	0.297	0.527
Total	1.000	1.000	1.000	1.000

- *Only the husband works* less often than the average population (29.7% compared with 52.7%).

- *Both parents work equally* more often than the average population (33.4% compared with 15.1%).

This opposition between profiles is the most important aspect (as highlighted by the first dimension) of the contingency table's deviation from independence or, furthermore, from the relationship between the two variables.

This aspect involves both extreme categories (as could very well be expected) and the average categories which play a more neutral role in the opposition. More generally, in the plane, the category *part-time work* is extremely close to the centre of gravity, thus indicating a profile near to the average profile (this can be observed directly on the table and calculated by this category's contribution to χ^2: 4.78%; see Table 2.3). This can also be expressed in the following manner: the women who answered *part-time work* cannot be distinguished from the rest of the population (in terms of their responses to question 1). This observation also suggests that the answer *part-time work* was chosen, at least in part, for that which Tabard describes as its "moderate nature" (especially for those who responded *only the husband works*). However, in the end, this answer does not seem to be particularly informative: when a woman gives this answer, this does not provide us with any further insight into what her answer to question 1 might be (technically, the conditional distribution of *part-time work* is equal to the marginal distribution). This sheds light on the contradiction between the impressions given by the two questions (though it must be noted that the answers to the questions about family imply that the participants' opinion of women's work is less favourable than the answers to the other question).

In synthetical terms, it may be said that the first dimension organises the categories of the second variable from the least favourable to women's work to the most favourable. As in PCA, dimensions are generally designated by one (or a few) words which sum up their meanings: here, it therefore seems natural to name this dimension "attitude to women's work." The word "attitude" must here be understood in psychological terms, that is, each object (here the concept of women's work) has a valence (either positive or negative); the opinions relating to the object are then organised according to that valence, and thus in a primarily one-dimensional manner. The attitude (of an individual toward an object) is represented by her position along this dimension.

2.3.4.2   Row Representation (Partner's Work)

The first dimension organises the categories from the least favourable to women's work (*only the husband works*) to the most favourable (*both parents work equally*). Here, again, we can call this dimension "attitude to women's work," and this is no coincidence. We can illustrate this arrangement by referring to the row profiles: we will leave the reader to do so, as we have already illustrated it in the example for the columns. We will, however, comment that the mediating category is at a distance from the origin of the axes (unlike for the cloud of columns), but remains clearly on the favourable side of women's work.

2.3.5   Superimposed Representation of Both Rows and Columns

Up until now, we have considered the cloud of rows N_I in $\mathbb{R}^J$ and the cloud of columns N_J in $\mathbb{R}^I$ separately. Each of these clouds was projected onto its directions of maximum inertia, projections which have been commented on separately, each with its own optimality (with each maximizing projected inertia). However, in CA, as in PCA, the analyses of both the cloud of rows and the cloud of columns are closely linked due to their relations of duality. Duality, or dual nature, stems from the fact that we are analysing the same data table but from different points of view (either rows or columns). This duality is particularly apparent and fruitful in CA as the rows and columns in contingency tables are fundamentally the same, that is to say, categories of categorical variables.

The first relationship has already been presented: clouds N_I and N_J have the same total inertia. In CA, the clear and indeed crucial nature of this total inertia (Φ^2 = deviation from independence) effectively illustrates that we are studying the same thing through either N_I or N_J.

The second relationship suggests that when projected onto the dimension of rank s (u_s for N_I in $\mathbb{R}^J$; v_s for N_J in $\mathbb{R}^I$), the inertia of N_I is equal to that of N_J and is denoted λ_s. Thus

$$\sum_{i=1}^{I} f_{i\bullet} \left(OH_i^s\right)^2 = \sum_{j=1}^{J} f_{\bullet j} \left(OH_j^s\right)^2 = \lambda_s.$$

So, not only do N_I and N_J have the same total inertia but they also have the same projected inertia on the dimensions of the same rank.

The third relationship, and that which is key to interpretation, brings the rows' coordinates and those of the columns together on the axes of the same rank. Thus

$$F_s(i) = \frac{1}{\sqrt{\lambda_s}} \sum_{j=1}^{J} \frac{f_{ij}}{f_{i\bullet}} G_s(j),$$

$$G_s(j) = \frac{1}{\sqrt{\lambda_s}} \sum_{i=1}^{I} \frac{f_{ij}}{f_{\bullet j}} F_s(i),$$

where $F_s(i)$ is the coordinate of the row profile i on the dimension of rank s (in $\mathbb{R}^J$); $G_s(j)$ is the coordinate of the column profile j on the dimension of rank s (in $\mathbb{R}^I$); and λ_s is the inertia of N_I (and N_J, respectively) projected onto the dimension of rank s in $\mathbb{R}^J$ (and in $\mathbb{R}^I$, respectively).

This property is fundamental to the superimposed representation, or "simultaneous representation," of rows and columns (see Figure 2.5, overlapping of the graphs in Figure 2.4). Thus for dimension s of this superimposed representation, up to the multiplicative factor $1/\sqrt{\lambda_s}$:

- A row i is at the barycentre of the columns, each column j having a weight $f_{ij}/f_{i\bullet}$ (these weights are positive and sum to 1).

- A column j is at the barycentre of the rows, with each row i having a weight $f_{ij}/f_{\bullet j}$ (these weights are also positive and sum to 1).

This property, which is referred to as barycentric (or sometimes pseudo-barycentric in order to evoke the coefficient $1/\sqrt{\lambda_s}$; also known as transition relations as they are used to transit from one space — $\mathbb{R}^I$ or $\mathbb{R}^J$ — to another) is used both to interpret the position of one row in relation to all of the columns and the position of one column in relation to all of the rows. Each row (or column, respectively) is near to the columns (and rows, respectively) with which it is the most closely linked, and is far from the columns (and rows, respectively) with which it is the least closely linked. Thus, in the example:

- *Stay at home* is on the same side as *only the husband works*, a category with which it is strongly related, and is on the opposite side from the other two categories, with which it is weakly associated.

- *Both parents work equally* is on the same side as *full-time work* and is at the opposite side to *stay at home*.

It must be noted that the origin of the axes coincides with the average profile (= barycentre) of each of the two clouds. Thus, when a row profile i has a positive coordinate:

- It is generally more associated to the categories j that have positive coordinates than it would have been in the independence model.

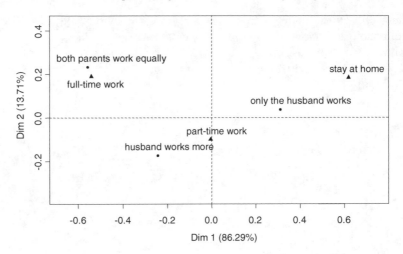

FIGURE 2.5
Simultaneous representation of rows and columns (= superimposed graphs
from Figure 2.4).

- It is generally less associated to the categories j that have negative coordinates than it would have been in the independence model.

The word "generally" in the previous sentence is important. Row profile
coordinates are determined by all of the column coordinates. We can therefore
comment on the position of a row with respect to the positions of all of the
columns but it is impossible to draw conclusions about the distance between
a specific row and a specific column. Associations between rows and columns
on which we wish to comment have to be checked directly from the data (the
contingency table).

Barycentre and Pseudo-Barycentre
During interpretation, the coefficient $1/\sqrt{\lambda_s}$ must always be taken into account. It is important to remember that, in CA, eigenvalues lie always between
0 and 1 (this will be explained in further detail later on). As a consequence,
when compared with exact barycentres, representations provided by CA are
expanded. Thus, in the example, $1/\sqrt{\lambda_1} = 2.93$ and $1/\sqrt{\lambda_2} = 7.33$:

- The category *stay at home* (in column), which is almost exclusively associated with the category *only the husband works* (in row), might almost be
  confused with the latter in an exact barycentric representation; its position
  within the plane is much further from the origin.

- The category *both parents work equally* (in row) is associated more or less
  equally (142 and 106) with the categories *part-time* and *full-time* and, in an
  exact barycentric representation, would be situated around half-way between

these two categories. On the plane, the category (in row) is much further from the origin and, along dimension 1, appears (slightly) beyond *full-time work*.

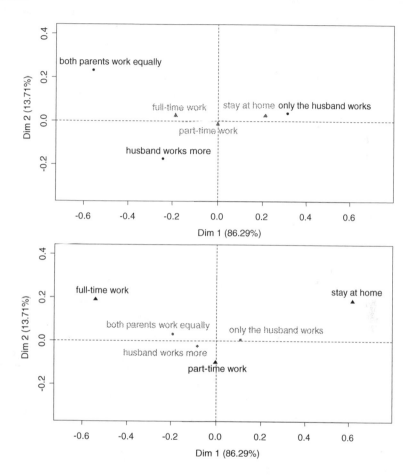

FIGURE 2.6
Representation of exact barycentres; the top graph represents the rows and the bottom the columns for a CA of Table 2.1.

One may be led to think that it could be more appropriate to represent the exact barycentres rather than the pseudo-barycentres. However, in this case, two graphs are necessary, and in each one the rows and the columns do not play symmetrical roles. Furthermore, the set of rows and the set of columns do not have the same inertia as the cloud of barycentres is centred around this origin (as compared to usual representation), which makes interpreting the associations between categories more difficult (see Figure 2.6).

The aim of an exact barycentric representation is to be able to visualise the

intensity of the relationship expressed by the plane (in terms of parts of Φ^2).
A cloud of barycentres (such as that of the rows) clustered around the origin
(along the dimension of rank s) illustrates a weak relationship (expressed
by the dimension of rank s) between the two variables V_1 and V_2 (each row
profile, being near to the origin, differs only slightly from the average profile).
However, in this case, the associations between the rows and columns are
difficult to see. This is what enables the expansion via the coefficient $1/\sqrt{\lambda_s}$,
an expansion which increases as the part of the relationship expressed by
the dimension weakens. It thus appears that the simultaneous representation
produced by the CA is designed to visualise the nature of the relationship
between the variables (i.e., relationships between rows and columns) and tells
us nothing about its intensity. This intensity is calculated from eigenvalues
which are components of Φ^2. In standard CA practice, both aspects of the
relationship between two variables (nature and intensity) are identified by
separate tools (graphs and eigenvalues).

Another advantage of quasi-barycentric representation can be observed
in the synthetic interpretation of the simultaneous representation in this example. The first dimension opposes the categories which are favourable to
women's work to those which are not. More precisely, the first dimension organises the categories of the two variables from the least favourable to women's
work (*stay at home*) to the most favourable (*both parents work equally*). With
this in mind, CA implies that *stay at home* is a much less favourable response
to women's work than *only the husband works*. This result tells us how the participants interpreted the multiple choices associated with the two questions.
We should therefore focus the search within the data to identify the origin
of the difference between the two categories perceived by CA. The category
which is furthest from the origin, *stay at home*, corresponds to the biggest
difference from the average profile, as demonstrated by its contribution to χ^2
(118.07 for *stay at home*; 88.34 for *only the husband works*). In more tangible terms, it may observed that almost all (84.9%) of those who replied *stay
at home* also replied *only the husband works*: this data thus groups together
the two least favourable responses to women's work. Again, in only 26.5%
of cases did women who replied *only the husband works* give both of these
unfavourable answers. In this case, it could be said that *stay at home*, which
predisposes women to give a second unfavourable response to women's work,
is in itself less favourable to women's work than *only the husband works*.

We will not attempt to offer a psycho-sociological explanation for the find-
ings in this table. We will, however, retain the idea that, through simultaneous
representation, CA clearly and simply illustrates the characteristics of the table being analysed which are not necessarily apparent when simply looking
at the data presented in the table. This quality, which is already so apparent
when used on a data table as small as this, becomes much more obvious and
indeed more valuable as the size of the table increases.

2.4 Interpreting the Data

As in PCA, CA interpretation is based on inertias and graphical representations (i.e., the coordinates of the rows and the columns on the dimensions). Nonetheless, when interpreting the data, one occasionally feels the need for indicators to answer a few specific questions. Some of these issues are listed below, along with a few tips on how to deal with them.

2.4.1 Inertias Associated with the Dimensions (Eigenvalues)

One of the results of the dual barycentric properties is another important attribute of CA, as introduced by the following reasoning:

1. We consider the projection of N_I on the dimension of rank s.

2. N_J is placed at the exact barycentres. Cloud N_J is thus "inside" N_I, which cannot be a cloud of barycentres of N_J.

3. The dual property, "N_I to the barycentres of N_J and N_J to the barycentres of N_I" is never true unless an expansion coefficient is considered. This coefficient must expand the cloud of exact barycentres and thus must be positive. Hence $\lambda_s \leq 1$.

In the particular case when $\lambda_s = 1$, the cloud N_I having been positioned, the only reason cloud N_J would not be located within cloud N_I (in terms of barycentres) would be in the case of a mutually exclusive association between rows and columns. Figure 2.7 illustrates the structure of data in this case: the set I of lines (and columns, J, respectively) can be divided into two subsets I_1 and I_2 (J_1 and J_2, respectively); I_1 (I_2, respectively) is exclusively associated with J_1 (J_2, respectively). This data structure indicates a strong relationship between the two variables V_1 and V_2, as illustrated by CA with a dimension moving apart, on the one hand I_1 and J_1, and on the other hand I_2 and J_2.

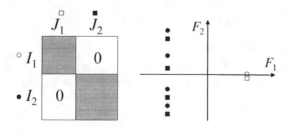

FIGURE 2.7
Case of an eigenvalue of 1; data structure and map defined by the two first dimensions ($\lambda_1 = 1$).

In practice, the eigenvalues of a CA are almost never exactly 1; however, a high eigenvalue indicates a structure similar to that in Figure 2.7, and it is essential to appreciate this when analysing contingency tables. It is therefore important to have a look at the eigenvalues in CA. In the example, the eigenvalues are rather small (see Table 2.5). Despite being associated with a strong structure, the first eigenvalue is small: therefore, that which is illustrated here is merely a tendency, even if it is highly significant (see χ^2 test).

Again, we will not attempt to propose a psycho-social interpretation for the "weakness" of this relationship. Might it simply be due to the fact that the questions are not identical? Or perhaps because of the "noise" which always encumbers the answers to opinion polls?

TABLE 2.5
Eigenvalues (= Projected Inertias) Resulting from a CA of Table 2.1

	Eigenvalue	Percentage of variance	Cumulative percentage
Dim 1	0.117	86.29	86.29
Dim 2	0.019	13.71	100.00

Having insisted on the fact that the inertia associated with a particular dimension is part of the relationship between the two variables V_1 and V_2, it seems natural to express this dimension as a percentage (see Table 2.5). In the example, it therefore appears that the first dimension represents 86.29%, and thus almost all of the difference between the actual sample (the data table) and the theoretical sample under the independence hypothesis. This is one of the arguments for considering this dimension alone for interpretation. More generally, eigenvalues calculate the relative importance of the dimensions: their order suggests the dimensions that we should focus on. With this in mind, we shall represent this sequence using a bar chart. Figure 2.8 illustrates a historic case (12 prevailing tobacco brands) (Benzécri, 1973, p. 339) in which five dimensions are proposed, each slightly larger than the others. The fact that the slow decrease of eigenvalues continues beyond the fifth suggests that the corresponding dimensions represent only "noise."

In a more detailed study of a case such as this, it is advisable to examine the sixth dimension, at least superficially, as a clear understanding of this dimension would mean that one would have to retain it when commenting the results. Although it might seem unusual, this common practice (taking into account interpretable dimensions even if they have only very weak inertias) is not a bad idea (it is difficult not to comment on a dimension that one knows how to interpret). However, it is the cause of a great deal of debate.

As the axes are orthogonal, the projected inertias from multiple axes can be added. In the example, 100% of the relationship is expressed by the plane. This is not a feature of the data but rather results from the dimensions of the table (3×3; see comment on the number of dimensions in Section 2.3.3). In more general terms, we use the sum of S first percentages of inertia to

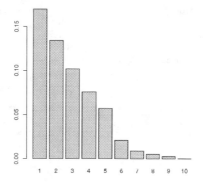

FIGURE 2.8
Example: bar chart illustrating the order of the eigenvalues of a CA.

measure the component of inertia accounted for in a commentary of the first dimensions S.

Returning to the idea of geometric interpretation of the eigenvalues in terms of projected inertia, the percentage of the inertia associated with the dimension is expressed as

$$\frac{\text{projected inertia of } N_I \text{ (or } N_J\text{) on the dimension of rank } s}{\text{total inertia of } N_I \text{ (or } N_J\text{)}} \times 100.$$

This criterion here appears as a measurement of the overall representation quality of cloud N_I (or N_J) by the dimension of rank s. More generally, we can examine projection on a plane. In this case, the criterion is a response to the following question: When projecting cloud N_I (or N_J) onto a plane (usually the first plane, constructed from dimensions 1 and 2), the cloud is distorted as the projection procedure can only reduce the distance between points. Is this distortion significant? In other words, do the distances between points (of one set, either rows or columns) on a plane successfully reflect the distances in the initial space ($\mathbb{R}^J$ or $\mathbb{R}^I$)? If they do, interpretation is simple, as the distances on the plane can easily be identified within the data even if eigenvalues are low. If they do not, the general value of the representation is not in question. However, poor representation quality indicates that other phenomena, visible on the following planes, are added to that shown by the plane in question. In such a case, it would be difficult to identify results shown by the analysis within the data, particularly in the case of low eigenvalues (but it is always easy when the eigenvalues have a value approaching 1).

Finally, the representation quality associated with a plane is one characteristic to be considered; however, it is not an absolute indicator about the interest of the plane. Our small example is a good illustration of this (despite being a limited case): the representation quality of 100% is due to the small dimensions of the table and in no way helps to predict the utility of the analysis.

Comment on the Maximum Value of Φ^2

The table $I \times J$ generates a maximum of $\inf(I-1, J-1)$ non-null eigenvalues, each of which is less than or equal to 1. The maximum value of Φ^2 is therefore $\inf(I-1, J-1)$. By relating the observed value of Φ^2 to its theoretical maximum, we are led to the statistical indicator known as Cramer's V, defined as

$$V = \left(\frac{\Phi^2}{\inf(I-1, J-1)} \right)^{1/2}.$$

The purpose of this criterion is to vary between 0 (independence) and 1 (maximum relationship in terms of each category of the variable with the greatest number of categories being exclusively associated with one category from the other variable). Due to its range of variation, Cramer's V acts in a similar manner as a correlation coefficient, up to a certain point. Thus, when presented with multiple categorical variables (defined on the same individuals), we can produce a V matrix (in the same way as we might construct a correlation matrix).

2.4.2 Contribution of Points to a Dimension's Inertia

The inertia associated with a dimension can be decomposed by points. The contribution of point i to the inertia of the dimension of rank s is generally expressed by (working from the notations in Section 2.3.3)

$$\text{ctr}_s(i) = \frac{\text{inertia of } i \text{ projected on the dimension of rank } s}{\text{inertia of } N_I \text{ projected on the dimension of rank } s},$$

$$= \frac{f_{i\bullet} \left(OH_i^s \right)^2}{\sum_{i=1}^{I} f_{i\bullet} \left(OH_i^s \right)^2} = \frac{f_{i\bullet} \left(OH_i^s \right)^2}{\lambda_s}.$$

This contribution is often multiplied by 100 or 1000 to facilitate the construction of the tables. It is often said to be "relative" as it is related to the entire cloud. "Absolute" contribution is thus the projected inertia of point $\left(f_{i\bullet} \left(OH_i^s \right)^2 \right)$. This distinction between "relative" and "absolute" contribution is not always referred to as such by all authors. Often, "contribution" (or even "absolute contribution") means that which is referred to in this book as "relative contribution."

Contributions are calculated for both rows and columns (Table 2.6). They may be added for multiple rows (respectively, columns). They are particularly useful when there are a great number of points. Choosing the points which contribute the most often facilitates interpretation. The atypical case of a dimension caused by only one or two points can thus be detected immediately: interpretation can therefore focus on this point(s) and thus prevent risky generalisations.

Due to the small sample size, the analysis of data on attitudes towards women's work does not necessarily require the use of contributions. However, the data are sufficient to be able to illustrate how contributions are calculated. For example, the contributions of *only the husband works* and *both parents work equally* on the first dimension demonstrate the respective roles of weights and distances when comparing two close contributions.

$$\text{ctr}_1(\text{only the husband ...}) \quad = \quad \frac{0.5267 \times 0.3096^2}{0.1168} = \frac{0.5267 \times 0.0958}{0.1168} = 0.432$$

$$\text{ctr}_1(\text{both parents ...}) \quad = \quad \frac{0.1514 \times 0.5586^2}{0.1168} = \frac{0.1514 \times 0.312}{0.1168} = 0.404$$

The point *both parents work equally* is (roughly) twice as far from the origin as the other, thus suggesting a greater influence. However, the weight of *both parents work equally* is (roughly) three times less, therefore simultaneously suggesting a lesser influence. As far as inertia is concerned (criterion used to define the dimensions), it is the square of the distance which intervenes: thus, in the end, both contributions balance out.

TABLE 2.6
Coordinates, Relative Contributions (in %), and Representation Quality for Each Category and Dimension

	Coordinates		Contributions		Representation Quality	
	Dim 1	Dim 2	Dim 1	Dim 2	Dim 1	Dim 2
both work equally	-0.56	0.23	40.43	44.43	0.85	0.15
husband works more	-0.24	-0.17	16.37	51.44	0.67	0.33
only husband works	0.31	0.04	43.20	4.13	0.99	0.01
stay at home	0.62	0.18	53.91	29.61	0.92	0.08
part-time work	0.00	-0.10	0.01	34.85	0.00	1.00
full-time work	-0.54	0.19	46.08	35.53	0.89	0.11

Remark

In CA, the points generally have different weights and the contribution calculations play a more important role than in standardised PCA (in which the elements have equal weights). Indeed, in standardised PCA contribution is proportional to the square of the distance to the origin and can be read (approximately) on the graphical representations.

2.4.3 Representation Quality of Points on a Dimension or Plane

The percentage of inertia associated with a dimension has been presented, among other things, as an indicator of the quality of a cloud's representation. This indicator can be applied to one point and we can thus calculate the representation quality of a point i by the dimension of rank s, which we denote $\text{qlt}_s(i)$ (see Section 1.6.1.2). Thus

$$\text{qlt}_s(i) \quad = \quad \frac{\text{inertia of } i \text{ projected on the dimension of rank } s}{\text{total inertia of } i}$$

$$= \frac{(OH_i^s)^2}{(Oi)^2} = \cos^2(\overrightarrow{Oi}, \overrightarrow{OH_i^s}).$$

This ratio indicates how the deviation of category i from the average profile is expressed on the dimension of rank s. Again, this indicator is not really pertinent for the results of CA as applied to the table of attitudes towards women's work. This is due to the small size of the table, which leads to a perfect representation of the clouds (and therefore each point) on the first (and only) plane. However, this data enables us to clearly illustrate the meaning of the indicator, as explained below:

1. The four extreme categories are successfully represented by the first dimension (representation quality: > 0.85). The deviation of each category from the average profile (i.e., the categories which are more or less connected than in the case of independence) is also successfully illustrated by this dimension. The other dimension tells us relatively little about these categories.

2. The category *part-time work* is ineffectively represented by the first dimension, but this does not necessarily mean that we should disregard it during interpretation (quite the reverse in fact; the central position of this category has indeed been commented on at length). This effectively illustrates the preeminence of the coordinates in the interpretation. In other words, the category's deviation from the average profile can only be perceived on other axes.

In practice, representation qualities are mainly used in the following cases:

• When examining one specific category; the representation quality enables us to select the plane on which the category is expressed the most successfully.

• When looking for a small number of categories to illustrate the significance of dimension s with the help of raw data. This is extremely useful for communicating results; first, the categories with the most extreme coordinates are selected (as the effect represented by dimension s is extreme). One must then decide between these categories in order to favour those that are represented most successfully (as the effect of dimension s is found only here).

It must be noted that these comments, made within the context of CA, can also be applied to other principal component methods (for example, by replacing the notion of the average profile in CA by the "average individual" for PCA).

2.4.4 Distance and Inertia in the Initial Space

Independently, or more precisely, prior to the results of CA, one might wonder which categories are the most, or indeed the least, "responsible" for the deviation from independence. Two different perspectives may be assumed:

1. Inertia has already been used in the χ^2 decomposition into rows or columns. Table 2.7 thus illustrates the relatively similar roles played by each of the four most extreme categories.

2. The distance from the average profile. In this instance, we do not take the category's sample size into account: the distances are summarised in Table 2.7, which gives the distance from the origin for each of the two categories *only the husband works* and *the husband works more* (the limited number of lines reduces the relevance of this indicator during interpretation and, in this respect, the dominant category *only the husband works* (52.7%) cannot differ greatly from the average profile as it is part of this average profile).

In practice, the distances from the origin are used to select the row and/or column which is the most, or least, similar to the average profile. This is a helpful way of illustrating the diversity of the profiles.

TABLE 2.7
Distance (Squared) to Average Profile and Inertia (in Initial Spaces $\mathbb{R}^I$ and $\mathbb{R}^J$)

	Both work equally	Husband works more	Only husband works
Distance	0.3665	0.0891	0.0973
Inertia	0.0555	0.0287	0.0512
	Stay at home	Part-time work	Full-time work
Distance	0.4158	0.0099	0.3287
Inertia	0.0685	0.0065	0.0604

2.5   Supplementary Elements (= Illustrative)

As with any principal component method (see PCA, Section 1.6.2), we can introduce supplementary elements (i.e., rows and/or columns), a term which refers to their status: they do not participate in the construction of the dimensions (however, they may still be projected onto the dimensions in the same way as the other active elements). They are also referred to as "illustrative" elements, in reference to the way in which they are the most frequently used; that is to say, to enrich and illustrate dimension interpretation.

In CA, supplementary elements are generally contingency tables. Their position within the plane is calculated using the barycentric properties. It must be noted that the expansion coefficient $1/\sqrt{\lambda_s}$ is dependent on the relationship between the active variables V_1 and V_2 and not the supplementary elements. The result is that the representation of a contingency table's categories introduced in additional columns (confronting V_1 with a third variable

V_3, for example) accounts for the strength of the relationship between V_1 and V_2. Thus, the cloud of the categories from V_3 (see Figure 2.9) will be more (or less, respectively) closely concentrated around the origin than that for the categories from V_1 if the relationship (and more precisely the part of the relationship expressed by the dimension in question) between V_1 and V_3 is less (or more, respectively) intense than that between V_1 and V_2. One might be led to believe that using another expansion coefficient for the supplementary elements would enable us to "better" visualise the associations between additional columns and active rows, for example. However this is not the case as, if we did so, we could no longer compare the relative positions of active and additional columns.

This can be illustrated by the following example: in his work, Tabard.[3] also published another table confronting V_1 and a new question (see Table 2.8), which we shall call V_3. This new variable is in a standard format for opinion polls. The participants are given a list of possible opinions: for each opinion, participants express agreement or disagreement using a scale, here a four-point scale from *totally disagree* to *totally agree*. The precise wording of the question is *What do you think of the following opinion: Women who don't work feel like they're cut off from the world?*

1. Totally agree

2. Somewhat agree

3. Somewhat disagree

4. Totally disagree

It must be noted that the relationship between V_1 and V_3 is highly significant ($\chi^2 = 162.19$; p-value $= 2.04 \times 10^{-32}$) but moderate in intensity ($\Phi^2 = 0.094$; $V(V_1, V_3) = 0.217$); in particular, it is less intense than the relationship between V_1 and V_2 ($\Phi^2(V_1, V_2) = 0.135$; $V(V_1, V_2) = 0.260$).

Going beyond the face value of the questions, this weak relationship refers to the "noise" which accompanies the responses to questions like V_3. A common problem with the categories expressing an agreement is that they may not oppose one another: those expressing disagreement may simply be expressing general hostility towards the questionnaire. This may be one of the reasons behind the contradictory responses which can conceal relationships between the questions.

We shall limit our commentary of the projection of the categories for variable V_3 to the following points:

- The categories expressing agreement with the opinion *women who stay at home feel cut off from the world* are found on the side of the unfavourable attitudes towards women's work, and inversely for the categories expressing disagreement. Interpretation of the dimension is therefore enhanced.

---

[3]Tabard N. (1974) *The needs and hopes of families and young people*, CREDOC, Paris.

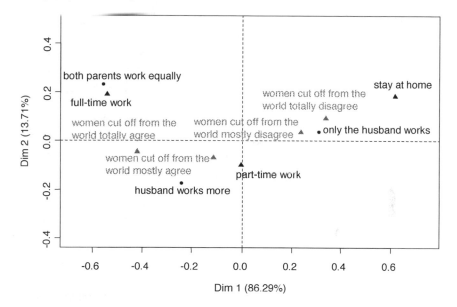

FIGURE 2.9
Representation of Figure 2.5 enhanced by the categories of the supplementary variable *women who stay at home feel cut off from the world*.

- The cloud of categories for V_3 is more tightly clustered around the origin than those of the other two variables. Here, we reencounter the fact that the relationship between V_1 and V_3 is less intense than that between V_1 and V_2.

- The category *totally agree* is further from the origin of the axes than *totally disagree*; it would therefore seem to be more typical of a favourable attitude towards women's work than *totally disagree* is typical of an unfavourable attitude.

Comment: Fields of Application for Correspondence Analyses
Correspondence analysis was designed to process contingency tables and its entire theoretical justification lies within this context. Nevertheless, using a CA program can provide useful results for many other types of tables containing positive numbers and whose margins can be interpreted. One example of such a use can be found in the case of an incidence matrix associated with a graph (whose general term x_{ij} is worth 1 if an edge connects the vertices i and j or 0 otherwise).

To justify the use of CA with a table such as this, and thus to be able to interpret the output, users must question the significance of the main properties of CA. Thus, within the case of an incidence matrix: (1) the barycentric property ensures that each vertex appears at the barycentre of those with which

TABLE 2.8
Table of Opinions Laid Out in Supplementary Columns of the CA for
Table 2.1

Women who don't work feel like they're cut off from the world?	Ideal family is the one in which:			Sum
	Both parents work equally	Husband works more	Only the husband works	
Totally agree	107	192	140	439
Somewhat agree	75	175	215	465
Somewhat disagree	40	100	254	394
Totally disagree	39	88	299	426
Sum	261	555	908	1724

they are connected via an edge; and (2) the maximum inertia ensures relevance, particularly in the first plane. It is as close as possible to the vertices linked by multiple paths of length 2 and separates the others.

2.6   Implementation with **FactoMineR**

In this section we shall demonstrate how to conduct a CA with FactoMineR. We will therefore again be working with the different results from the CA of Table 2.1, which have already been commented on in the previous sections.

```
> library(FactoMineR)
> work <- read.table("http://factominer.free.fr/bookV2/work_women.csv",
    header=TRUE,row.names=1,sep=";")
> summary(work)
```

The χ^2 test and Table 2.2 are obtained using only the first three columns of the dataset:

```
> res.test.chi2 <- chisq.test(work[,1:3])
> res.test.chi2
> round(res.test.chi2$expected,1)
```

Table 2.3 is obtained using

```
> round(res.test.chi2$residuals^2, 2)
> round(100*res.test.chi2$residuals^2/res.test.chi2$stat,2)
```

Having transformed the data table into a matrix, Table 2.4 is obtained by

```
> dd <- rbind(work,apply(work[,1:3],2,sum))
> rownames(dd)[4] <- "Mean profile"
> round(prop.table(as.matrix(dd),margin=1),3)

> dd <- cbind(work,apply(work[,1:3],1,sum))
> colnames(dd)[4] <- "Mean profile"
> round(prop.table(as.matrix(dd),margin=2),3)
```

The CA is then performed. By default it provides a graph depicting the superimposed representation (see Figure 2.5).

```
> res.ca <- CA(work[,1:3])
```

The graph representing the rows and that representing the columns (see Figure 2.4) are obtained using the function **plot.CA**.

```
> plot(res.ca,invisible="col")
> plot(res.ca,invisible="row")
```

The graphs representing the exact barycentres (see Figure 2.6) are obtained using

```
> plot(res.ca,invisible="col")
> coord.col = sweep(res.ca$col$coord,2,sqrt(res.ca$eig[,1]),FUN="*")
> points(coord.col,pch=17,col="red")
> text(coord.col,rownames(coord.col),col="red")

> plot(res.ca,invisible="row")
> coord.row = sweep(res.ca$row$coord,2,sqrt(res.ca$eig[,1]),FUN="*")
> points(coord.row,pch=20,col="blue")
> text(coord.row,rownames(coord.row),col="blue")
```

The main results (inertia associated with each component, coordinates, contributions, and representation qualities of the rows and columns) are obtained with the function **summary.CA**:

```
> summary(res.ca)

Call:
CA(X = work[, 1:3])

The chi square of independence between the two variables is equal
          to 233.4304 (p-value =  2.410248e-49).

Eigenvalues
                          Dim.1    Dim.2
Variance                  0.117    0.019
% of var.                86.292   13.708
Cumulative % of var.     86.292  100.000

Rows
                              Iner*1000   Dim.1   ctr  cos2   Dim.2   ctr  cos2
Both parents work equally |      55.49 | -0.56 40.43  0.85 |  0.23 44.43  0.15 |
Husband works more        |      28.68 | -0.24 16.37  0.67 | -0.17 51.44  0.33 |
Only the husband works    |      51.24 |  0.31 43.20  0.99 |  0.04  4.13  0.01 |

Columns
                              Iner*1000   Dim.1   ctr  cos2   Dim.2   ctr  cos2
stay.at.home              |      68.49 |  0.62 53.91  0.92 |  0.18 29.61  0.08 |
part.time.work            |       6.48 |  0.00  0.01  0.00 | -0.10 34.85  1.00 |
full.time.work            |      60.43 | -0.54 46.08  0.89 |  0.19 35.53  0.11 |
```

The graph of eigenvalues is obtained with

```
> barplot(res.ca$eig[,1],main="Eigenvalues",
    names.arg=1:nrow(res.ca$eig))
```

The inertias of rows and columns (see Table 2.7) are also given by the function **summary.CA** whereas the distances squared must be recalculated using the row margin and the column margin:

```
> res.ca$row$inertia/res.ca$call$marge.row
> res.ca$col$inertia/res.ca$call$marge.col
```

The graph in Figure 2.9 is obtained by conducting a new CA, specifying that, after column 4, all of the subsequent columns are supplementary:

```
> res.ca2 <- CA(work,col.sup=4:ncol(work))
```

2.7 CA and Textual Data Processing

The methodologies which focus on analysing a group of texts in terms of the words contained within them are grouped together under the term "textual data." At the heart of these methodologies, we find correspondence analysis of a given table (known as the lexical table or the occurrence table) of texts and words. Generally, x_{ij} denotes the number of times the word j is used in the text i. At first glance, it seems to be simply a different use of the standard methods of data analysis. However, textual data have a number of unique features, requiring specific methodologies. From this perspective, it becomes clear that we are dealing with a specific and separate scientific domain (which also has its own conferences: for example, JADT — *Journées d'analyse des données textuelles* or Textual Data Analysis Days). We shall first examine this data as a new field of application and then as a scientific domain.

Let us go back to the occurrence table mentioned above. It may be perceived as a contingency table (and therefore use a CA) when considered in the following manner. The most basic statistical unit is the graphical form, a sequence of characters positioned according to two separators (punctuation marks and space). Each graph is characterised by two categorical variables: the variable "text" (whose categories are the texts themselves), and the variable "dictionary" (whose categories are words). The occurrence table distributes graphical forms according to these two variables and, in this respect, is also a contingency table.

CA is well suited to studying this kind of table (and indeed it is for precisely this purpose that it was invented: the first ever application of CA to be published, in Brigitte Escofier's thesis in 1965,[4] was of this type); more precisely, it describes the deviation between this table and the independence

---

[4]Escofier B. (1965) *L'analyse des correspondances*, PhD, University of Rennes, France.

model. CA's approach to this deviation is based on the notion of profile: in this context, we are referring to a text's lexical profile (the overall frequencies of the words in a text) and a word's usage profile (overall frequency of this word in the texts).

Independence occurs when all of the profiles (both usage and lexical) are identical, and thus the same as the average profile (total number of words in each text; overall usage frequency of each word). The deviation from independence increases as these profiles differ from the average profile. CA analyses this deviation in order to summarise the information in the form of a visualisation organised as a sequence of dimensions. A dimension may, for example, identify a group of texts which all have a high frequency (i.e., higher than the average profile) for some words and a low frequency (i.e., lower than the average profile) for others. Through duality, this same dimension may also identify a group of words which share the property of having a high frequency (i.e., higher than the average profile) for certain texts. These are the words which characterise the texts identified by this same dimension. Thus, the visualisation provided by CA corresponds exactly to that which we might expect to achieve through the exploratory analysis of a set of texts.

The distinguishing feature of textual data appears in the way the table is constructed, i.e., the choice of the rows and of the columns.

Which Texts Can Be Used?

Until now, for convenience, we have referred to "text" as that found in a row in an occurrence table. Defining these texts is not always simple, as we shall illustrate in the following two examples.

In the first application of CA, the original corpus is taken from the play *Phèdre* by Racine, first performed in 1677. To analyse this body of text, it must first be broken down. The criterion chosen was that of character: one row of the table (and therefore one text) contains all of the lines for a given character. It therefore became possible to construct a map of the characters in terms of the vocabulary that they use. The first dimension corresponds to social status: great men (here, Phèdre, although generalisations are tempting) do not use the same words as their subjects (in French, this is most obviously characterised by the use of the words "tu" and "vous"). There are other ways of dividing up the text: by scene (to picture the sequence of events) or, more specifically, by confronting characters and acts, in order to follow the evolution of the characters over the course of the play.

One primary application of textual data analysis is to study open questions in questionnaires. One famous example of this type of question is the following pair of questions, asked consecutively: for you, what is the Right? For you, what is the Left? The purpose of this type of questions is (almost) universally recognised: the spontaneity of the answers is proof of the importance the participant places on the aspects involved; information which can otherwise be difficult to obtain. In this example, do we obtain economic, social, or political aspects? Are these aspects the same for the Left and the Right?

First, we might want to consider each participant as a row of the table.

However, this table is somewhat sparse (many cells are empty), and analysing it using CA is often tedious and in the end rather disappointing (many dimensions identify small groups of individuals with a few specific words in common) as it is not very synthetical. One recommended methodology is to group together the surveys according to a criterion which confronts all or some of the following traditional variables: gender (male/female), educational level, and age (broken down into several classes). There are, of course, other ways of grouping the data, and the user should think carefully about this choice, as it can have a strong effect on the results. A text is therefore an amalgamation of the answers to one of the categories resulting from the way the participants are grouped together.

Which Words?

Again, to simplify matters, we have until now referred to "word" as that found in a column of an occurrence table. In practice, the definition of that which a column should represent is rather tricky as there are many different opinions on the matter, each with interesting aspects. The user must make those decisions which seem to him/her to be the most suited to his/her data and objectives. Some of the key considerations are outlined below.

Selection According to Overall Frequency. Unusual words are both uninteresting for the user in synthetical terms, and can also modify the CA. Indeed, a word used in only one text containing only this word can create a dimension associated to the (maximum) eigenvalue of 1. Even if this never actually happens in practice, it is a useful case of reference, taking into account high eigenvalues of around 0.5 which are frequently observed in this type of analysis. We will therefore eliminate the rare words from the data. The notion of rarity is of course relative and the threshold must be defined for each case depending on the overall frequency of words.

Tool Words. Should articles, conjunctions, and so forth, be retained for analysis? At first glance, users are tempted to eliminate these words, which seem insignificant. It must be noted that if these words are evenly spread throughout the text according to the average profile (when their occurrence does not depend on the length of the text), they are therefore close to the cloud's centre of gravity and have little effect on the analysis. However, if their usage frequency is dependent on the text, they indicate something and thus should be retained.

Lemmatisation. Should we, for example, group together the graphical representations corresponding to the singular and the plural of the same noun? Or to the same verb? Lemmatisation consists of grouping together the graphs relating to one dictionary entry. There is both support and opposition to this approach. Let us show the property of distributional equivalence, a traditionally general property of CA which was notably highlighted using lexical tables illustrated by a "textual" example: if *day* and *days* have the same profile, it makes no difference whether they are considered together or separately. This property is one of the strong arguments against lemmatisation: if the profiles are equal, nothing is gained; however, if they are not, that variation is lost.

Nonetheless, in practice, one must ensure that the above variation really merits our attention, which is usually not the case for the rarest of words (unless we use an exaggeratedly high selection threshold).

Stemmatisation. This technique consists of grouping together the graphs with the same root. Thus, in comments made during a wine tasting, we might want to group together *lively* and *young*. Stemmatisation has similar inconveniences to lemmatisation; however, here the risk of confusing distinct notions is much higher.

Repeated Segments. Some words tend to appear in combination with other words and this combination ("repeated segments") is often much more meaningful than the words considered separately and can help avoid ambiguity. Thus, remaining in the field of wine tasting, *red fruits* is much more evocative than *fruit* (think of the taste of *dried fruits* in sweet wines) and *red* (the colour *red* does not, in principle, imply the flavour qualities of *red fruits*). The best example of repeated segments has to be *social security*, the meaning of which cannot be easily inferred from either *social* or *security*. It is therefore highly useful to consider repeated segments by assigning each one a column.

The above considerations are by no means exhaustive in textual data analysis but offer the key points of reference for its implementation. Thus, the majority of the work involved is required prior to conducting the CA, in order to establish the lexical table from a body of texts.

The package tm ("text mining") is dedicated to the analysis of textual data. FactoMineR's **textual** function is a lexical function from which a contingency table can be constructed. Let us illustrate this function with the following example, which contains two categorical variables and one textual variable:

```
> work
      Millésime    Wine              Text
1 Millésime 1 Wine 1  Acidity, fruity
2 Millésime 2 Wine 1    Fruity, light
3 Millésime 1 Wine 1            Woody
4 Millésime 2 Wine 1          Acidity
5 Millésime 1 Wine 2            Sweet
6 Millésime 2 Wine 2    Sweet, syrupy
7 Millésime 1 Wine 2    Light, fruity
8 Millésime 2 Wine 2     Sweet, light
```

The **textual** function enables us to construct the contingency table for each category of one or more categorical variables and/or for each combination of categories of the two categorical variables. The argument sep.word defines the word separators and the argument maj.in.min transforms all of the words to lower case letters. The following command line builds a contingency table with the words as columns and the second variables and category combinations from the first and second variables in the rows. It also gives the number of times each word is used (nb.words) and the number of rows in which it features (an output which is useful for texts but irrelevant for open questions as single words are never repeated).

```
> textual(don,num.text=3,contingence.by=list(2,1:2),sep.word=", ",maj.in.min=TRUE)
$cont.table
```

	acidity	fruity	light	sweet	syrupy	woody
Wine 1	2	2	1	0	0	1
Wine 2	0	1	2	3	1	0
Millésime 1.Wine 1	1	1	0	0	0	1
Millésime 1.Wine 2	0	1	1	1	0	0
Millésime 2.Wine 1	1	1	1	0	0	0
Millésime 2.Wine 2	0	0	1	2	1	0

```
$nb.words
```

	words	nb.list
sweet	3	3
light	3	3
fruity	3	3
acidity	2	2
woody	1	1
syrupy	1	1

2.8   Example: The Olympic Games Dataset

2.8.1   Data Description — Issues

The data table confronts, in the rows, the different athletic events and, in the columns, the different countries. Each cell contains the total number of medals (gold, silver, and bronze) won at the Olympiads between 1992 and 2008 (Barcelona 1992, Atlanta 1996, Sydney 2000, Athens 2004, Beijing 2008). An extract of the dataset is provided in Table 2.9. Over the 5 Olympiads, 58 countries won at least 1 medal in one of the 24 events: 10 000 m, 100 m, 110 m hurdles, 1 500 m, 200 m, 20 km, 3 000 m steeplechase, 400 m, 400 m hurdles, 4×100 m, 4×400 m, 5 000 m, 50 km, 800 m, decathlon, discus, high jump, javelin, long jump, marathon, hammer, pole jump, triple jump. The table contains a lot of zeros as only 360 medals were awarded whereas there are 1392 cells in the table.

The data is available in the FactoMineR package:

```
> library(FactoMineR)
> data(JO)
```

This is indeed a contingency table, with the 360 medals as individuals. Two categorical variables are associated with each medal: the event to which it refers and the country which won that medal. The table confronts these two variables.

From a rather formal perspective, in a table such as this we are interested in studying the relationship between the two variables: event and country. However, this expression is not particularly pertinent. It may be rendered more tangible in the following way: by looking for an association between the

TABLE 2.9
Olympic Data: Number of Medals Won by Event and by Country over 5
Olympiads (the 10 countries that won the most medals)

	USA	KEN	RUS	GBR	ETH	CUB	MAR	GER	JAM	POL
10 000 m	0	4	0	0	8	0	2	0	0	0
100 m	5	0	0	1	0	0	0	0	1	0
110 mH	9	0	0	0	0	3	0	1	0	0
1500 m	0	5	0	0	0	0	3	0	0	0
200 m	8	0	0	1	0	0	0	0	1	0
20 km	0	0	3	0	0	0	0	0	0	1
3000 m Steeple	0	12	0	0	0	0	1	0	0	0
400 m	11	1	0	1	0	0	0	0	1	0
400 mH	7	0	0	1	0	0	0	0	2	0
4x100 m	4	0	0	1	0	2	0	0	1	0
4x400 m	5	0	1	2	0	1	0	0	2	0
5000 m	0	5	0	0	4	0	3	1	0	0
50 km	0	0	4	0	0	0	0	1	0	3
800 m	1	5	1	0	0	0	0	1	0	0
Decathlon	5	0	0	0	0	1	0	1	0	0
Discus	0	0	0	0	0	1	0	3	0	1
High Jump	3	0	3	2	0	2	0	0	0	1
Javelin	0	0	2	3	0	0	0	0	0	0
Long Jump	7	0	0	0	0	2	0	0	1	0
Marathon	1	3	0	0	3	0	1	1	0	0
Hammer	1	0	0	0	0	0	0	0	0	1
Pole Jump	4	0	3	0	0	0	0	1	0	0
Shot Put	8	0	0	0	0	0	0	0	0	1
Triple Jump	3	0	2	3	0	2	0	0	0	0

events and countries which might be considered remarkable in one way (i.e.,
a given country only wins medals in one specific event) or another (a given
country wins no medals in a specific event and thus compensates by winning
many in another event).

Resorting to the notion of profile, which is indeed the cornerstone of CA,
is here, as is often the case, more natural, more accurate, and more beneficial.
First, a country's athletic profile is defined by looking at all the medals, broken
down into events (one column from the table). The question then becomes:
can each country be considered to have the same athletic profile or, on the
contrary, do some countries perform better in certain events? If this is the
case, can these "specialisations" be summarised? For example, by highlighting
the oppositions between, on the one hand, the countries with the same profiles
(i.e., who performed well in the same events) and, on the other hand, those
which have opposing profiles (i.e., do not perform well in the same events).

Similarly, the way in which the medals for one event are distributed defines
the "geographical profile" for that event (one row from the table). Can each
event be considered to have the same geographical profile or, on the contrary,
are some events the speciality of specific countries? Might we be able to
summarise these specialisations by highlighting the oppositions between, on
the one hand, the events which share the same profile (i.e., won by the same

countries) and, on the other hand, the events with opposing profiles (i.e., won by other countries)?

The two approaches outlined above are dependent on the notion of similarity between the profiles. In this similarity, each country's total number of medals should not be considered, as it would mean dividing those countries which obtained many medals from the others, in which case CA would no longer be useful. Also, the above notion of profile must be considered in terms of CA, that is to say, in terms of conditional probability or, more simply, percentages (of medals obtained by each country for each event).

Comment: Margins

The very nature of this data means that the sum of the column (or column margin) must be constant and equal to: 3 (types of medals) × 5 (Olympiads)= 15 (although there are some rare exceptions due to medals being withdrawn). This has two consequences. First, in the analysis, the events have the same weight (and thus, with a constant profile, they have the same influence). Second, the "average" athletic profile, which acts as a point of reference (located at the origin of the axes) is therefore a constant profile. CA, which highlights the deviations from the average profile, will indicate the important role played by the countries with a very specific athletic profile (the most extreme case of which would be if one country won all its medals in the same event).

The row margin contains the total number of medals won by each country. These figures vary greatly (from 1 medal for 18 countries to 82 medals for the USA). Country weights therefore vary a great deal: with constant profiles, the countries with the most medals have a greater influence on the analysis. The average profile, located at the origin of the axes, therefore contains the proportions of medals won by the countries (very different to a constant profile). Thus, an event A may be more characteristic of a country X than of a country Y even if X won more medals than Y in this event (because Y won more medals than X overall).

The margins may be calculated after the correspondence analysis has been conducted (see the end of this section for how to obtain the margins).

2.8.2   Implementation of the Analysis

Here, we consider all rows and lines as active. To conduct this analysis, we use the **CA** function of the package FactoMineR. The primary input parameters are the data table, the indexes of the supplementary rows, and the indexes of the supplementary columns. By default, none of the rows or columns are considered supplementary (`row.sup=NULL` and `col.sup=NULL`), in other words, all of the elements are active.

```
> res.ca <- CA(JO)
```

The **CA** function provides the CA graph as well as the numerical output which can be summarised by the **summary.PCA** function. It must be specified that these numerical indicators are rounded to 2 decimal places, nb.dec=2, and that the results are provided by the first 10 elements of each table nbelements=10 (10 is the value by default), in order to keep the text brief. To obtain all of the elements for each table, we would use nbelements=Inf.

```
> summary(res.ca, nb.dec=2)

Call:
CA(X = JO)

The chi square of independence between the two variables is
            equal to 2122.231 (p-value =  2.320981e-41 ).

Eigenvalues
                        Dim.1   Dim.2   Dim.3   Dim.4   Dim.5   Dim.6   Dim.7   Dim.8   Dim.9
Variance                 0.82    0.62    0.54    0.48    0.40    0.36    0.33    0.32    0.27
% of var.               13.85   10.53    9.23    8.16    6.72    6.17    5.55    5.35    4.56
Cumulative % of var.    13.85   24.38   33.62   41.78   48.50   54.67   60.23   65.58   70.14
                        Dim.10 Dim.11 Dim.12 Dim.13 Dim.14 Dim.15 Dim.16 Dim.17 Dim.18
Variance                 0.24    0.23    0.16    0.16    0.14    0.13    0.12    0.10    0.09
% of var.                4.16    3.91    3.11    2.78    2.46    2.22    2.06    1.76    1.58
Cumulative % of var.    74.29   78.20   81.31   84.09   86.55   88.77   90.82   92.58   94.16
                        Dim.19 Dim.20 Dim.21 Dim.22 Dim.23
Variance                 0.08    0.08    0.07    0.06    0.04
% of var.                1.44    1.35    1.27    1.05    0.73
Cumulative % of var.    95.60   96.95   98.21   99.27  100.00

Rows (the 10 first)
            Iner*1000 Dim.1    ctr   cos2  Dim.2    ctr   cos2  Dim.3    ctr   cos2
10000m   |    366.59 | -2.16  23.85  0.53 |-0.33   0.73   0.01 |-0.17   0.23   0.00 |
100m     |    262.73 |  0.68   2.35  0.07 |-1.16   9.09   0.21 |-0.41   1.27   0.03 |
110mH    |    157.04 |  0.59   1.80  0.09 |-0.50   1.66   0.07 |-0.40   1.20   0.04 |
1500m    |    338.17 | -1.47  11.02  0.27 |-0.18   0.23   0.00 | 0.37   1.07   0.02 |
200m     |    200.00 |  0.72   2.61  0.11 |-1.08   7.89   0.24 |-0.47   1.68   0.05 |
20km     |    330.79 |  0.28   0.41  0.01 | 1.04   7.21   0.14 | 1.48  16.68   0.27 |
3000mSt  |    271.03 | -1.61  13.23  0.40 |-0.15   0.15   0.00 | 0.13   0.12   0.00 |
400m     |    137.13 |  0.48   1.18  0.07 |-0.74   3.64   0.16 |-0.31   0.75   0.03 |
400mH    |    265.58 |  0.53   1.44  0.04 |-0.79   4.14   0.10 |-0.41   1.26   0.03 |
4x100m   |    202.24 |  0.55   1.54  0.06 |-0.65   2.87   0.09 |-0.40   1.20   0.03 |

Columns (the 10 first)
            Iner*1000 Dim.1    ctr   cos2  Dim.2    ctr   cos2  Dim.3    ctr   cos2
alg      |     55.56 | -1.00   1.35  0.20 |-0.10   0.02   0.00 | 0.34   0.24   0.02 |
aus      |     72.22 |  0.45   0.41  0.05 | 0.59   0.95   0.08 | 0.96   2.82   0.21 |
bah      |    102.78 |  0.69   0.49  0.04 |-0.64   0.56   0.03 | 0.44   0.30   0.02 |
bar      |     63.89 |  0.75   0.19  0.02 |-1.48   0.98   0.09 |-0.55   0.16   0.01 |
bdi      |     63.89 | -2.07   1.45  0.19 |-0.24   0.03   0.00 |-0.12   0.01   0.00 |
blr      |     94.44 |  0.42   0.36  0.03 | 1.63   7.17   0.47 |-1.38   5.80   0.33 |
bra      |    102.78 | -0.02   0.00  0.00 |-0.54   0.40   0.02 |-0.52   0.41   0.02 |
brn      |     63.89 | -1.63   0.90  0.11 |-0.23   0.02   0.00 | 0.51   0.13   0.01 |
can      |     55.56 |  0.58   0.46  0.07 |-0.41   0.29   0.03 | 0.09   0.02   0.00 |
chn      |     63.89 |  0.66   0.15  0.02 |-0.63   0.18   0.02 |-0.54   0.15   0.01 |
```

Prior to the CA, the χ^2 test indicated whether or not the table's deviation from independence might be caused by "random fluctuations" (as, unlike CA, this test accounts for the overall sample size). χ^2 has a value of 2122 and is associated with a p-value of 2.32×10^{-41}.

Here, however, the total sample size ($5 \times 5 \times 24 = 360$ medals) is extremely small in terms of the number of cells in the table ($24 \times 58 = 1392$). We are thus far from the conditions of test validity (even the most "relaxed" of which consider 80% of the theoretical sample size to be greater than 5, and the others to be greater than 1) and the p-value here can be considered for information only. However, the p-value is here so small that the significance of the deviation of this table from independence cannot be questioned.

2.8.2.1   Choosing the Number of Dimensions to Examine

As in all principal component methods, the study of the inertia of the dimensions enables us to see whether or not the data is structured, and, on the other hand, to determine the number of dimensions to interpret.

The first table given by the function **summary.CA** contains the eigenvalue (i.e., the inertia or explained variance) associated with each dimension, the percentage of inertia it represents in the analysis, and the sum of these percentages. These eigenvalues can be visualised using a bar chart (see Figure 2.10):

```
> barplot(res.ca$eig[,1],main="Eigenvalues",
    names.arg=paste("dim",1:nrow(res.ca$eig)))
```

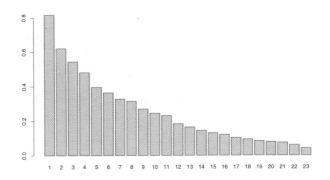

FIGURE 2.10
Olympic data: eigenvalues associated with each dimension of the CA.

The first two dimensions express 24.40% of the total inertia. It may be interesting to consider the next dimensions, which also express a high percentage of the total inertia.

2.8.2.2   Studying the Superimposed Representation

The superimposed representation of the CA (see Figure 2.11) is a default output of the function **CA**.

We can identify all of the row coordinates (and column, respectively) in the function `res.ca$row` (`res.ca$col`, respectively). We therefore obtain a table detailing the coordinates, contributions (indicating to what extent an

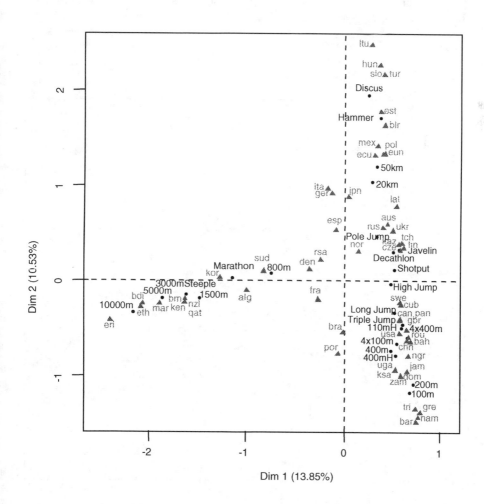

FIGURE 2.11
Olympic data: superimposed representation.

individual contributes to the construction of a dimension), cosines squared (measuring the quality of the projection of individuals on a dimension), and the inertias for each element (corresponding to the distance from the barycentre counterbalanced by the weight of the element).

A graph can also be constructed for dimensions 3 and 4 (see Figure 2.12). In this case we use the function **plot.CA** (which can be summoned using **plot** or **plot.CA**). We then specify the axes of representation (axes = 3:4):

```
> plot(res.ca,axes=3:4)
```

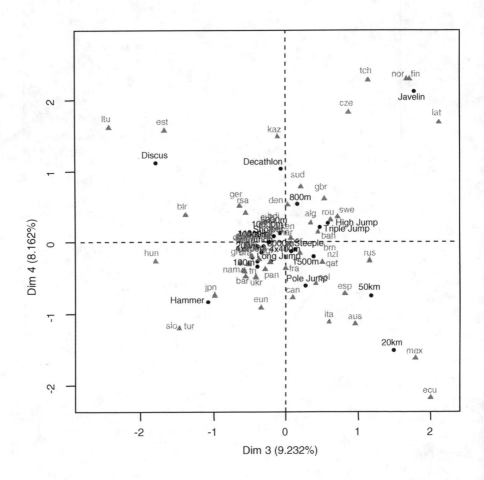

FIGURE 2.12
Olympic data: superimposed representation on plane (3, 4).

2.8.2.3 Interpreting the Results

First, let us examine the projections of the various events on the plane induced by the first two dimensions. The results here are rather astonishing, as all of the events involving long-distance races are separated from the other events on the first dimension. Among other things, there is a gradient of the different events going from 10 000 m to 800 m. All of the events, without exception, are sorted from the longest distance to the shortest. This means that the results of the 10 000 m are more noteworthy than those of the other endurance events. Nonetheless, the marathon is much closer to the centre of the graph than expected. This is due to the fact that it is not the same kind of endurance event as the others.

The countries with negative coordinates on the first dimension are those which won many medals in the endurance events compared with the results of these countries in the other events, but also compared with the number of medals won by the other countries in endurance events. Here we find those African countries that specialise in endurance events (Eritrea, Ethiopia, Burundi, Morocco, Qatar, Kenya) and New Zealand (N.B., New Zealand only won one medal, in the 1 500 m, which is why it has such an extreme coordinate).

It is interesting here to examine the contributions of the different countries. It must be noted that in CA, unlike PCA, the most extreme elements are not necessarily those which contributed the most to the construction of the dimensions, as the weights differ from one element to another. The contributions of the 13 countries which contributed most to the construction of dimension 1 are featured below (countries are organised in descending order):

```
> res.ca$col$contrib[rev(order(res.ca$col$contrib[,1])),1]
    ken     eth     mor     usa     gbr     eri     cub     bdi     alg     jam     tri
 31.387  22.072  12.160   9.149   2.139   1.947   1.683   1.452   1.352   1.313   1.119
```

Ethiopia, Kenya, and Morocco account for 65% of the construction of the first dimension. These countries won a great number of medals in these events. The three countries won a total of 60 medals, of which 59 were obtained in endurance events.

The second dimension separates sprinting events from the discus, the hammer, and walking (20 km and 50 km). Again, here, there is a gradient between the sprinting events: the 100 m is more extreme than the 200 m and the 400 m. The relay events are also less extreme than the individual events. Here, we can clearly observe that the 400 m is a sprinting event, whereas the 800 m is an endurance event. In the same way, walking (20 km and 50 km) is distinguishable from both sprinting and endurance events. And again, the 50 km is more extreme than the 20 km.

The countries which won medals in the sprinting events are Barbados, Namibia, Trinidad and Tobago, Jamaica, the Dominican Republic, and so forth. The contributions of these countries to the construction of the second dimension are much more consistent than for the first dimension. The USA contributed significantly to the construction of this dimension even though its coordinate is relatively close to 0. This is due to the great number of medals won by this country: a total of 82, of which 49 were won in sprinting events (compared to the percentage of sprinting events: 7/24). The 15 countries which contributed most to the construction of dimension 2 are detailed below:

```
> res.ca$col$contrib[rev(order(res.ca$col$contrib[,2])),2]
   usa    ltu    blr    hun    pol    eun    tri    est    ger    nam    jam
11.324 10.942  7.175  6.911  6.314  5.582  4.790  4.234  3.766  3.643  3.629
```

For the discus and hammer events, eastern European countries such as Lithuania, Hungary, Slovenia, and, also Turkey won the most medals.

Dimensions 3 and 4 also separate the discus and the hammer from the walking events (20 km and 50 km). The javelin is a very different event than those of the hammer and the discus. Northern European countries (Norway, Czechoslovakia, the Czech Republic, Finland, and Latvia) score high in the javelin event.

The row and column margins can be obtained thus (as can the number of medals obtained by each country, by multiplying the column margin by the total number of medals, 360):

```
> res.ca$call$marge.row
> res.ca$call$marge.col[rev(order(res.ca$call$marge.col))]
  usa   ken   rus   eth   gbr   cub   ger   mar   jam   pol   esp   ita
0.228 0.097 0.053 0.042 0.042 0.039 0.028 0.028 0.025 0.022 0.022 0.019
> res.ca$call$marge.col[rev(order(res.ca$call$marge.col))]*360
usa ken rus eth gbr cub ger mar jam pol esp ita
 82  35  19  15  15  14  10  10   9   8   8   7
```

2.8.2.4 Comments on the Data

Athletics fans might at first be disappointed when reading this example. Indeed, CA first overturns the biggest trends emanating from this data: the endurance events are dominated by African athletes, sprinting events are dominated by the USA, and sprint, endurance, and throwing events are relatively diverse. However, this is exactly what we ask of an analysis: to identify the primary characteristics.

Nevertheless, when we look closer, certain results are interesting and may attract the curiosity of athletics fans (or even athletes themselves!). We shall outline a few of these results and leave the specialists to interpret them as they see fit.

- The results of the CA point to a distinct separation between endurance events (1 500 m, 3 000 m steeple, 5 000 m, 10 000 m, and marathon) and sprinting events (100 m and 200 m). The two resistance events, the 400 m and the 800 m, are not included and the other running events are separated into two categories with a limit situated somewhere between the 400 m and the 800 m (the 800 m is similar to the endurance events whereas the 400 m is similar to the sprinting events). Energy is therefore managed differently in these two events.

- Furthermore, the marathon is indeed an endurance event but does not behave in the same way as the others: it is much more extreme than may have been expected. Similarly, the walking events (20 km and 50 km) are not endurance events in the same way as running races.

- The athletes who participate in sprinting events often tend to "double up" and run 100 m and 200 m or 200 m and 400 m. The graph shows that 100 m and 200 m are very similar events, and are more similar than 200 m and 400 m. The 100 m and 200 m are both power events whereas the 400 m is a resistance event.

- The two hurdle events (110 m and 400 m hurdles) are somewhat different: the 110 m hurdles is relatively far from the 100 m whereas the 400 m hurdles is close to the 400 m. The 100 m and 110 m hurdles events draw on very different qualities: the 110 m hurdles is a technical event whereas the 100 m requires bursts of energy, which explains why no athletes participate in both of these events. However, the 400 m hurdles is a much less technical event than the 110 m hurdles. It draws on qualities similar to those used in the 400 m, which explains why athletes can participate in both events.

- In the throwing events, the hammer and the discus are very similar whereas the shot put and the javelin are somewhat different. The hammer and the discus are thrown in rotation (with a leverage effect) while the javelin is thrown in a straight line. The shot put is thrown with or without rotation (and without leverage, as the shot must be in contact with the neck when thrown).

- The decathlon, the ultimate all-round event, is opposed to the endurance events on the first dimension. This event is therefore not suitable for endurance athletes, generally speaking. Decathlon athletes' morphology, with a great deal of muscle mass and a capacity for bursts of energy, does not help them in endurance events. Decathlon athletes often have difficulty finishing their event by a 1 500 m.

It must be noted that all comments are made entirely on the basis of the number of medals won by different countries and by event, and in no way reflect the physical capabilities of the athletes who participated in these events.

2.9 Example: The White Wines Dataset

2.9.1 Data Description — Issues

Within the context of research into the characteristics of the wines from Chenin vines in the Loire Valley, a set of 10 dry white wines from Touraine were studied: five Touraine Protected Appellation of Origin (AOC) from Sauvignon vines, and five Vouvray AOC from Chenin vines (see Table 2.10).

These wines were chosen with the objective of illustrating the diversity of Loire Valley wines within each specific vine (although we can imagine that, given the profession of the person responsible for choosing the wine, there may be one restriction to this diversity: he most likely did not choose wines that he does not like). It must be noted that there is very little connection between the Protected Appellation of Origin (AOC) and the vine (in terms of experimental planning). For this reason, and to simplify matters, we shall limit ourselves to the vine.

TABLE 2.10
Wine Data: The 10 Wines Studied

Number	Name	Vine	Label of Origin	Comments
1	Michaud	Sauvignon	Touraine	
2	Renaudie	Sauvignon	Touraine	
3	Trotignon	Sauvignon	Touraine	
4	Buisse	Sauvignon	Touraine	
5	BuisseCristal	Sauvignon	Touraine	
6	AubSilex	Chenin	Vouvray	7 g of residual sugars
7	Aub.Marigny	Chenin	Vouvray	Cask-aged
8	FontDomaine	Chenin	Vouvray	
9	FontBrûlés	Chenin	Vouvray	
10	FontCoteaux	Chenin	Vouvray	Cask-aged

These wines were used for a number of sensory analyses, combining different types of evaluations and protocols. The data analysed are from a wine tasting by 12 professionals and are of a "textual" nature. The instructions were: for each wine, give one (or more) word(s) which, in your opinion, characterises the sensory aspects of the wine. As is the tradition in wine tasting, the wines were presented "blind"; that is to say, those tasting the wines did not know which wines they were tasting. Nonetheless, as this tasting took place at the Trade Fair for Loire Valley Wines, the tasters could easily have guessed that they were tasting Loire Valley wines, even if it was never specified. However, given the diversity of Loire Valley wines in terms of soil, vines, and wine-making techniques, it may be assumed that the very general knowledge about the entire set of wines had no real effect on the relative characteristics of each one.

Participants were thus given a questionnaire comprising 10 open questions (one per wine). This data was brought together in a table with the wines as rows and the columns as words, where the general term x_{ij} is the number of times that a word j was associated with a wine i (see Table 2.11).

TABLE 2.11

Wine Data: The Number of Times Each Word Was Used for Each Wine (30 Words)

	1S-Mic	2S-Ren	3S-Tro	4S-Bui	5S-Bui	6C-Aub	7C-Aub	8C-Fon	9C-Fon	10C-Fon	Total
Fruity	1	5	5	3	4	0	1	4	3	1	27
Sweet, smooth, syrupy	0	1	1	0	0	11	1	2	1	1	18
Oak, woody	1	0	0	0	2	0	7	0	1	5	16
Light, supple	1	0	2	2	1	2	0	0	4	0	12
Acidic	1	0	1	2	1	0	2	1	2	1	11
Citrus	2	3	1	1	1	0	0	3	0	0	11
Golden yellow	2	0	0	1	0	1	2	1	2	2	11
Lively	2	3	0	1	3	1	1	0	0	0	11
Fruity flavours	2	1	2	1	0	1	0	1	1	0	9
Delicate, discrete	0	2	1	4	0	0	0	1	1	0	9
Bitter	1	1	0	0	0	0	0	1	2	3	8
Floral	0	1	2	0	2	0	0	1	1	1	8
Rich, musty	0	0	0	0	0	2	2	1	2	1	8
Pale yellow, pale	1	2	2	0	1	2	0	0	0	0	8
Fresh, cool	1	2	2	2	0	0	0	0	0	0	7
Long finish, persistant	1	1	1	0	0	0	2	0	1	1	7
Floral, white flowers	2	1	1	0	1	0	0	0	0	1	6
Dry	0	0	0	3	1	0	0	1	1	0	6
Intense, full	1	0	0	0	0	1	1	0	1	1	5
Honey	0	1	0	0	0	1	1	1	1	0	5
lack of character, weak	0	0	0	0	0	3	0	2	0	0	5
Open, lots of character	2	0	1	0	0	0	0	1	1	0	5
Full-flavoured	1	1	1	1	0	0	0	0	0	0	4
Foreign flavour (wax, tyre)	0	0	0	0	0	0	3	0	0	1	4
Vigourous flavour	2	0	2	0	0	0	0	0	0	0	4
Salty	1	1	0	1	1	0	0	0	0	0	4
Slightly acidic	1	0	0	1	2	0	0	0	0	0	4
Little character, expressivity	0	0	0	0	0	1	2	0	1	0	4
Sauvignon	1	1	1	0	0	0	0	0	0	1	4
Strong flavour	1	0	0	0	2	0	1	0	0	0	4
Total	28	27	26	23	22	26	26	21	26	20	245

This table can be viewed as a contingency table, considering that there are n sensory descriptions (a description is the association between a word and a wine), and that these descriptions are divided between two categorical variables: the wine to which it refers, and the word used. The CA will analyse the distance between the table and the independence model, a model in which each wine has the same word profile and, proportionally, each word is used the same number of times for each wine.

This type of table is often constructed and analysed using CA (historically, the first CA published was on an analogue table [5]), although usually the sample size tends to be much larger. Here we find ourselves very near to the threshold for CA use, as the total number of occurrences ($n = 245$) is extremely low. Nonetheless, the analysis is possible as wine experts tend to use rather standardised vocabulary, thus yielding a low total number of words

[5] Escofier B. (1965). *L'analyse des correspondances*, PhD, University of Rennes, France.

and thus a "sufficient" number of words with an acceptable frequency. Prior to the analysis, a number of "neighbouring" words were grouped together (for example, *sweet, smooth,* and *syrupy,* all of which refer to the same perception, that of the sweet taste of the wine). In this text, in order to simplify matters we use the term "word" for the rows in Table 2.11, even when they represent groups of words as they appear in the questionnaire (e.g., *vigourous flavour*) or when grouped together subsequently, for example, *sweet, smooth,* and *syrupy.* In this type of analysis, we eliminate the words which are used the least frequently. Due to the small sample size, the threshold below which words were no longer included was set at 4. This threshold was determined empirically: with this data, setting the threshold at 5 would not change the representation significantly, but would remove important words (such as "Sauvignon"). On the other hand, a threshold of 3 would lead to heavily laden graphs and words with weak coordinates.

The aim of this analysis is to provide an image summarising the diversity of the wines. Usually, the sensory diversity of wines is studied using a much stricter protocol: a questionnaire is established using a list of descriptors (acidity, bitterness, etc.), a jury is then asked to evaluate the wine using the descriptors, and then a final evaluation is conducted. One of the aims of our study is methodological; is it possible to obtain a significant image of the diversity of the wines using a simplified procedure (the tasters do not all participate at the same time, and they are free to use their own vocabulary)?

2.9.2   Margins

Examining the margins is important both in terms of direct interpretation (Which words are used the most often? Are some wines more prone to comments than others?), and in terms of their influence in CA (as a weight).

The word used most frequently is "fruity," which seems to correspond to our observations of comments about wine-tasting (if you are in any doubt, just look at the label on any bottle of wine, and you will almost always encounter the expression "wonderfully fruity"). The second most common is *sweet, smooth, syrupy.* It must be noted that these wines are all dry and thus any perception of sweetness would constitute a remarkable characteristic. Finally, the perception expressed by the term *oak,* associated with wines aged in barrels, is often used by professionals, and thus was frequently cited in this study (unlike an odour which might be perceived easily but which is unrecognisable and therefore expressed differently depending on who is describing it). Extending the commentaries of these sample sizes goes beyond the context of a data analysis book such as this. On a technical level, the weight of the words will increase with the frequency at which they are cited.

However, the number of words per wine seems to be stable. No one wine seems to attract significantly more comments than the other, which is undoubtedly a consequence (and indeed a desired result) of the form of instructions given ("for each wine, give one or more words ..."). To be sure, we can conduct a χ^2 test of best fit for the 10 observed samples (final row of Table 2.11) to uniform distribution. The p-value (0.97) confirms that we must not focus on the differences between the numbers of words for each wine. In terms of CA, we can consider that the wines will all have around the same influence on the analysis. When the number of words per wine differs, the analysis attributes a greater weight to a wine if it is the subject of a great number of commentaries (it has a more well-known profile).

2.9.3 Inertia

Total inertia (Φ^2) has a value of 1.5, and subsequently a χ^2 of 368.79 ($n = 245$) associated with a p-value equal to 1.23×10^{-5}. This table has poor validity test conditions (in theory, at least 80% of the theoretical sample size must have a value higher than 5 and none of them should be nil). However, the p-value is so low that the worth of CA for this data cannot be questioned. It must be noted, for those who are familiar with the diversity of Loire Valley wines (especially when we consider that these wines were chosen with the aim of illustrating this diversity), that the relationships between the words and the wines are to be expected. The question "put to the χ^2 test" is not so much that which proves the existence of a relationship but rather that which relates to the capacity of a set of data which is so limited that this relationship becomes evident. Here then, we may consider the response to be positive, but the data do not have the "statistical reliability" of the table of opinions on women's work (remember: for the latest table, p-value $= 10^{-49}$). We shall therefore be doubly careful when interpreting the data. In tangible terms, this means frequently referring back to the raw data and drawing connections with other external information. These comments are all the more important as there is no doubt about the significance of the CA, which only accounts for probabilities.

The intensity of the relationship, measured using Cramer's V, is rather strong: 0.409 (where 1 corresponds to a mutual exclusivity between each wine and a group of words, something that would be almost impossible in blind wine tasting). This value is higher, for example, than that of the data on women's work (0.26).

CA is implemented using the following commands:

```
> library(FactoMineR)
> data.wine = read.table("http://factominer.free.fr/bookV2/wine.csv",
    header=TRUE,row.names=1,sep=";",check.names=FALSE)
> res.ca=CA(data.wine,col.sup=11,row.sup=nrow(data.wine))
> barplot(res.ca$eig[,1],main="Eigenvalues",
    names.arg=1:nrow(res.ca$eig))
> summary(res.ca, nb.dec=2, ncp=2)

Call:
CA(X = wine, row.sup = nrow(wine), col.sup = 11)

The chi square of independence between the two variables is
```

equal to 368.7893 (p-value = 1.228089e-05).

Eigenvalues

	Dim.1	Dim.2	Dim.3	Dim.4	Dim.5	Dim.6	Dim.7
Variance	0.44	0.37	0.18	0.16	0.10	0.10	0.07
% of var.	28.93	24.67	12.05	10.35	6.65	6.35	4.38
Cumulative % of var.	28.93	53.60	65.65	76.00	82.65	89.00	93.38
	Dim.8	Dim.9					
Variance	0.06	0.04					
% of var.	4.13	2.49					
Cumulative % of var.	97.51	100.00					

Rows (the 10 first)

	Iner*1000	Dim.1	ctr	cos2	Dim.2	ctr	cos2
Fruity	| 46.77 |	-0.47	5.49	0.51 |	0.01	0.00	0.00 |
Sweet, smooth, syrupy	| 206.91 |	0.93	14.51	0.31 |	-1.34	35.69	0.64 |
Oak, woody	| 146.61 |	0.95	13.43	0.40 |	1.11	21.85	0.55 |
Light, supple	| 49.20 |	-0.12	0.15	0.01 |	-0.32	1.37	0.10 |
Acidic	| 18.63 |	0.03	0.01	0.00 |	0.37	1.64	0.33 |
Citrus	| 48.93 |	-0.63	4.09	0.36 |	-0.17	0.35	0.03 |
Golden yellow	| 26.02 |	0.45	2.11	0.35 |	0.28	0.93	0.13 |
Lively	| 46.53 |	-0.38	1.46	0.14 |	0.11	0.15	0.01 |
Fruity flavours	| 19.02 |	-0.38	1.25	0.29 |	-0.33	1.05	0.20 |
Delicate, discrete	| 70.86 |	-0.82	5.63	0.35 |	-0.17	0.28	0.01 |

Columns

	Iner*1000	Dim.1	ctr	cos2	Dim.2	ctr	cos2
1S.Michaud	| 109.34 |	-0.40	4.20	0.17 |	0.23	1.67	0.06 |
2S.Renaudie	| 102.92 |	-0.60	9.00	0.38 |	-0.10	0.32	0.01 |
3S.Trotignon	| 110.54 |	-0.59	8.50	0.34 |	-0.13	0.51	0.02 |
4S.Buisse	| 167.57 |	-0.81	14.15	0.37 |	-0.02	0.02	0.00 |
5S.BuisseCristal	| 143.58 |	-0.45	4.12	0.12 |	0.38	3.44	0.09 |
6C.AubSilex	| 309.20 |	0.94	21.33	0.30 |	-1.38	54.11	0.65 |
7C.Aub.Marigny	| 237.90 |	1.07	27.92	0.51 |	0.87	21.68	0.34 |
8C.FontDomaine	| 94.22 |	-0.05	0.04	0.00 |	-0.50	5.83	0.23 |
9C.FontBrls	| 86.91 |	0.22	1.14	0.06 |	0.03	0.02	0.00 |
10C.FontCoteaux	| 143.08 |	0.72	9.59	0.29 |	0.75	12.41	0.32 |

Supplementary row

	Dim.1	cos2	Dim.2	cos2
Total	| 0.00	10.73 |	0.00	4.47 |

Supplementary column

	Dim.1	cos2	Dim.2	cos2
Sum	| 0.00	2.12 |	0.00	0.67 |

The sequence of eigenvalues (see Figure 2.13 or the summary of the outputs given by the function **summary.CA**) shows two dimensions with noticeably higher inertias than on the following dimensions, which, when considered along with the accrued inertia percentage of 53.6%, leads us to focus our interpretation on the first plane. These two dimensions each have rather high inertia (0.4355 and 0.3713): associations between wines and words should therefore be clearly visible.

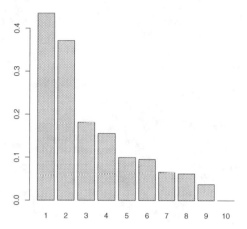

FIGURE 2.13
Wine data: chart of eigenvalues from the CA of Table 2.11.

2.9.4 Representation on the First Plane

Many different grids are available for analysis. For grids per axis we prefer, at least to start with, to use a grid per pole based on the wines. There are three possible poles (see Figure 2.14):

1. Aubuissières Silex (6), characterised by *sweet*, cited 11 times for this wine. This is the only wine to contain more than trace level residual sugars. This unusual (although authorised) characteristic for a dry wine, stands out, as it is only rarely cited for the other wines (7 times in total, but never more than twice for one wine), and accounts for over a third of the words associated with this wine. The graph highlights the wine's lack of character; although this term was only cited 3 times for this wine, we have classed it in second place (among other things, this characteristic is really a lack of a characteristic and is therefore less evocative).

2. Aubuissières Marigny (7) + Fontainerie Coteaux (10). These two wines were mainly characterised by the terms *oak, woody,* which were each cited 7 and 5 times, respectively, whereas the word was only used 3 times elsewhere. This description can, of course, be linked to the fact that these two wines are the only two to have been cask aged. According to this plane, *foreign flavour* best characterises these wines, but we chose to place it second due to the low frequency of this term (4), even if it was cited for these two wines alone. It should also be noted that the effect of ageing wine in casks does not only lead to positive characteristics.

3. The five Touraine wines (Sauvignon; 1–5). Characterising these

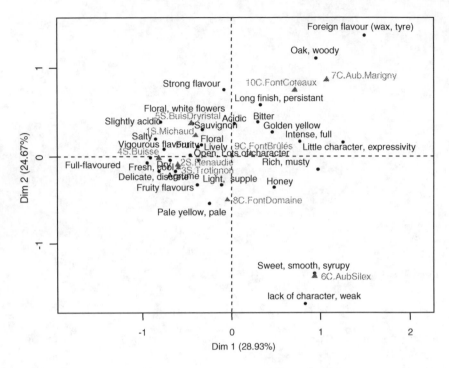

FIGURE 2.14
Wine data: first two dimensions map of the CA for Table 2.11.

wines was more difficult. The terms *lots of character*, *fresh*, *delicate*, *discrete*, and *citrus* were cited for these wines, which seems to fit with the traditional image of a Sauvignon wine, according to which this vine yields fresh, flavoursome wines. We can also add two more marginal characteristics: *musty* (and *little character*, respectively), cited 8 times (4 times, respectively), and which are never used to describe the Sauvignon wines.

Once these three poles are established, we can go on to qualify the dimensions. The first distinguishes the Sauvignons from the Chenin wines based on freshness and flavour. The second opposes the cask-aged Chenin wines (with an oak flavor) with that containing residual sugar (with a sweet flavour).

Having determined these outlines, the term *lack of character*, which was only used for wines 6 and 8, seems to appear in the right place, i.e., far from the wines which could be described as flavoursome, whether the flavour be due to the Sauvignon vines or from being aged in oak casks.

Finally, this plane offers an image of the Touraine white wines, according to which the Sauvignons are similar to one another and the Chenins are more varied. This could therefore be understood in a number of different, noncontradictory ways:

- There is only one way of producing Sauvignon whereas there are numerous ways of making Chenin.

- Viticulturists "work harder" to produce Chenin, the noble Touraine vine, by testing many different techniques in order to obtain the best flavour.

In saying these things, we are moving away from our role as statisticians, but we merely wanted to illustrate some of the ways in which the final user might interpret the results.

Dimensions 3 and 4

We shall consider the next dimensions only briefly, due to space restrictions. Following this approach, contributions may be useful to summarise a dimension.

Thus, from the point of view of contributions, dimension 3 confronts wines 1 and 4 with the words *dry delicate* and *vigorous*. These associations/oppositions are evident in the data. However, aside from the fact that they both have small sample sizes, they do not suggest that they require interpretation. Dimension 4 underlines wine number 5, which associates the words *lively* and *slightly acidic*. Again, this association can be observed in the data (just barely), but it also deals with small sample sizes and does not imply anything specific (on the contrary, *lively* and *slightly acidic* tend to oppose one another).

Conclusions

From a viticulturist's point of view, this analysis identifies the marginal characteristics of the Chenin vine. In practice, this vine yields rather varied wines which seem particularly different from the Sauvignons as they are somewhat similar and rather typical.

From a sensory standpoint, it is possible to obtain a reliable representation (here, reliability is suggested by clear relationships between the sensory "descriptions" and the "external" information available about the vine and whether or not the wine is cask-aged) with a much less strict procedure than usual (only one session).

From a statistical point of view, CA seems well-suited to analysing "marginal" matrices (with many either low-value or empty cells). However, the fact that we discarded those words with extremely low frequency (≤ 3) should not be overlooked.

2.10   Example: The Causes of Mortality Dataset

2.10.1   Data Description — Issues

This contingency table opposes causes of death, with age subdivided into age groups for the French population, for each year from 1979 to 2006. In each table (corresponding to a specific year), we can observe, at the intersection of the rows i and columns j, the number of individuals belonging to a given age group j who died (in the given year) due to cause i. To simplify matters, we mainly refer to the tables corresponding to the years 1979 and 2006, and their totals. We also refer to the table which opposes the French population for the entirety of the period from 1979 to 2006, for all causes, years, and age groups. The general term for this last table is, at the intersection of the rows i and columns j, the number of individuals belonging to a given age group j who died in year i (from all of the possible causes). These tables are grouped together in columns as shown in Figure 2.15.

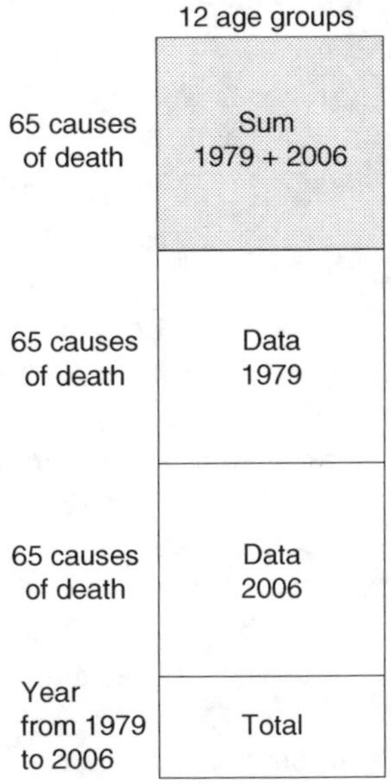

FIGURE 2.15
Mortality data: structure of the data table.

Data was taken from the *Centre d'épidémiologie sur les causes médicales de décès* (Cépidc) which offers access to some of its data online at http://www.cepidc.vesinet.inserm.fr/.

The real issue at hand here is to study the relationship between age and cause of death. Initially, the age variable is quantitative. The transformation of this variable, through dividing the range of variation into intervals, into a categorical variable, plainly highlights the nonlinear aspect of this relationship. This expectation of a nonlinear relationship stems from prior knowledge of the phenomenon being studied. This is particularly relevant when defining the age groups which are intended to group together individuals in a standardised manner, in terms of causes of mortality. The age groups were therefore sorted by periods of 10 years for the majority of the age range. However, as is often the case when dividing data into groups, the exceptions are almost all found at the extremities but here they carry rather different meanings: grouping together individuals of over 95 years of age means that the sample for this group remains an acceptable size. On the other hand, the youngest individuals are sorted into much smaller groups, as there is good reason to believe that newborns (0–1 year) and young children (1–5) may be subject to different causes of mortality than the other age groups.

By introducing the table grouping together all the deaths between 1979 and 2006 as active, we avoid focusing too much on one specific year and thus our results will have higher validity. With this in mind, it was possible to simultaneously analyse all of the years for the specified period, and not simply the two extreme years. This choice is purely practical, with the aim of avoiding a huge volume of data (whilst at the same time conserving maximum variability of the annual tables, which hypothesises a steady evolution over time).

The coordinates of the rows and columns on the dimensions of the CA for the active table provide a framework for analysing the relationship between age and cause of death for the specified period. Introducing these annual tables as supplementary rows means that the evolution of this relationship can be analysed within this framework, in terms of causes of mortality. This perspective can be justified as follows: each row in the active table, i.e., a cause of death, has a corresponding distribution of individuals who "belong" to this cause according to the age groups, which we refer to as the "age profile." The aim of the CA can be explained as a way of highlighting the main dimensions of variability for these profiles. For example, we expect a dimension which will oppose "young" profiles (causes of death representative of young people) with "elderly" profiles (causes of death characteristic of elderly people).

The supplementary rows are therefore also age profiles; they each correspond to a cause of death for a given year. Thus, there are many age profiles for each cause of death (or more specifically, there are many points on the graph), and it will be possible to analyse the evolution of these profiles using observations such as a given cause of death, which was typical for young people in 1979, was much less so in 2006.

2.10.2   Margins

The margins indicate the age groups which are more subject to mortality, and the most frequent causes of death. They also designate the weights of each category of the CA. The two margins vary greatly (see Figure 2.16 and Figure 2.17). The diagrams and the numerical data can be obtained using the following commands:

```
> library(FactoMineR)
> death <- read.table("http://factominer.free.fr/bookV2/death.csv",
    header=TRUE,sep=";",row.names=1)
> colnames(death) <- c("0-1","1-4","5-14","15-24","25-34","35-44",
    "45-54","55-64","65-74","75-84","85-94","95+")
> res.ca=CA(death,row.sup=66:nrow(death), graph=FALSE)
> round(res.ca$call$marge.col,3)
> round(res.ca$call$marge.row[order(res.ca$call$marge.row)],3)
> par(las=1)
> barplot(res.ca$call$marge.col,horiz=TRUE)
> barplot(res.ca$call$marge.row[order(res.ca$call$marge.row)],horiz=TRUE)
> par(las=0)
```

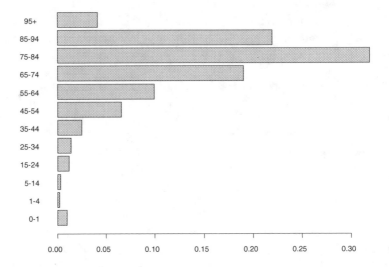

FIGURE 2.16
Mortality data: age group margins.

The most frequent cause of death is linked to cerebrovascular diseases. The age group with the highest number of deaths is the range between 75 and 84 years. The higher age groups (85–94 years and 95 years and over) feature fewer deaths simply because there are far fewer people of this age. It may also be noted that the number of deaths in the lowest age group (0–1 year) is relatively high when compared to the next age groups. This is even more

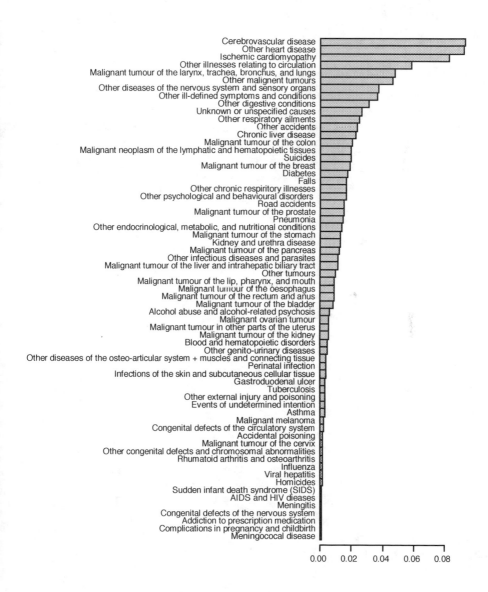

FIGURE 2.17
Mortality data: margins for causes of death.

surprising as this age group includes individuals all born in the same year, whereas the other groups include groups with a range of 4 and then 10 years. The percentage of children 0–1 year who die is therefore much higher than the percentage for children 1–4 years or 5–14 years.

2.10.3   Inertia

When applied to the active data, the χ^2 test of independence finds the relationship between the data to be significant. The observed χ^2 has a value of 1,080,254, and the associated p-value is extremely close to 0 (the software gives a result of 0). The significance of the text was expected due to that which we can all observe around us (even the simple fact that the expression "infant mortality" exists), and from the very high values observed in the data table. Although the test hypotheses are not validated (many of the cells have theoretical sample sizes of less than 5), the p-value is so low that the significance cannot be questioned. Total inertia is equal to $\Phi^2 = 1.0213$, and the intensity of the relationship, measured using Cramer's V, is high: 0.305 (1 would indicate an exclusive association between each age group and a group of causes of mortality).

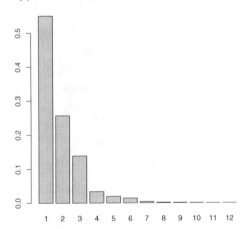

FIGURE 2.18
Mortality data: chart of eigenvalues.

```
> res.ca=CA(death,row.sup=66:nrow(death))
> summary(res.ca, nb.dec=4)
> barplot(res.ca$eig[,1],main="Eigenvalues",
    names.arg=1:nrow(res.ca$eig))
Eigenvalues
```

	Dim.1	Dim.2	Dim.3	Dim.4	Dim.5	Dim.6
Variance	0.5505	0.2570	0.1385	0.0338	0.0199	0.0143
% of var.	53.9002	25.1628	13.5653	3.3141	1.9439	1.4022
Cumulative % of var.	53.9002	79.0630	92.6283	95.9424	97.8863	99.2885

	Dim.7	Dim.8	Dim.9	Dim.10	Dim.11
Variance	0.0037	0.0017	0.0013	0.0004	0.0001
% of var.	0.3665	0.1624	0.1256	0.0439	0.0132
Cumulative % of var.	99.6550	99.8174	99.9430	99.9868	100.0000

The sequence of eigenvalues (see Figure 2.18) identifies three dimensions. These three dimensions account for 92.6% of the total inertia and therefore effectively summarise overall variability (within a $12 - 1 = 11$-dimensional space). We can therefore focus our interpretation on these first three dimensions.

Prior to conducting the CA, that is to say, with the filled spaces, it may be interesting to decompose the inertia by row and by column. The commands `res.ca$row$inertia` and `res.ca$col$inertia` carry the total inertia decomposed by row and by column. It is most fitting to express these inertias as percentages. For the columns, we obtain

```
> 100*res.ca$col$inertia/sum(res.ca$col$inertia)
  0-1   1-4  5-14 15-24 25-34 35-44 45-54 55-64 65-74 75-84 85-94   95+
52.62  2.16  1.67 12.22  6.18  3.99  4.56  3.97  2.08  2.39  5.34  2.82
```

The inertia for the age group 0–1 year is high, as 52.6% of the total inertia is due to this age group. "Half" of the relationship between age and cause of death therefore resides in the characteristic of this age group, which will therefore have a strong influence on the results of the CA. After this first set, the two other age groups which contribute the most to this relationship are 15–24 years and 25–34 years. These age ranges have very specific mortality profiles and will also strongly influence the CA.

For the 65 causes of death, we here list those with the five highest inertias (in the entire space), in descending order:

```
>100*res.ca$row$inertia[rev(order(res.ca$row$inertia))]/sum(res.ca$row$inertia)
                      Perinatal infection  32.41
                           Road accidents  13.70
                                     SIDS   7.94
Congenital defects of the circulatory system   6.54
                                 Suicides   5.00
```

Perinatal infection has high inertia compared with the other causes of mortality (32.41%); however, its weight is relatively low (with a margin of 0.00336). This cause of death has a very specific age profile (as suggested by the name).

By thoroughly inspecting the data, we can view the details of the calculations of these inertias in the form of a table summarising the weight (equal to the margin expressed as a percentage), the distance from the origin, and the inertia (raw, and as a percentage) for each row and each column. Thus, for the rows:

```
> bb<-round(cbind.data.frame(res.ca$call$marge.col,
    sqrt(res.ca$col$inertia/res.ca$call$marge.col),
    res.ca$col$inertia,res.ca$col$inertia/sum(res.ca$col$inertia)),4)
> colnames(bb)<-c("Weight","Distance","Inertia","% of inertia")
```

	Weight	Distance	Inertia	% of inertia
0-1	0.0099	7.3829	0.5374	0.5262
1-4	0.0021	3.2375	0.0221	0.0216
5-14	0.0032	2.3039	0.0170	0.0167
15-24	0.0118	3.2583	0.1248	0.1222
25-34	0.0140	2.1275	0.0632	0.0618
35-44	0.0251	1.2736	0.0408	0.0399
45-54	0.0657	0.8413	0.0465	0.0456
55-64	0.0994	0.6390	0.0406	0.0397
65-74	0.1900	0.3342	0.0212	0.0208
75-84	0.3189	0.2765	0.0244	0.0239
85-94	0.2189	0.4993	0.0546	0.0534
95 +	0.0410	0.8375	0.0288	0.0282

It would therefore seem that the strong contribution of the age group 15 to 24 years stems primarily from the distance from the origin, and therefore is a highly specific mortality profile.

2.10.4   First Dimension

The first dimension separates newborns of 0 to 1 years from the other age groups (see Figure 2.19). Figure 2.20 draws attention to the specific causes of mortality in this age group, that is to say, infant diseases which affect this age group exclusively, or quasi-exclusively (perinatal infection, SIDS, etc.). In this case, CA therefore highlights a specific phenomenon of a given category.

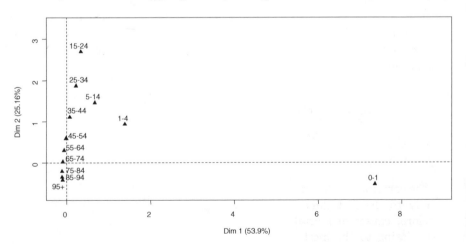

FIGURE 2.19
Mortality data: representation of age groups on the first plane.

In CA, as the elements (rows and columns) do not have the same weight, one must consider the contributions before proposing an interpretation. The commands `res$col$contrib` and `res$row$contrib` contain the contribu-

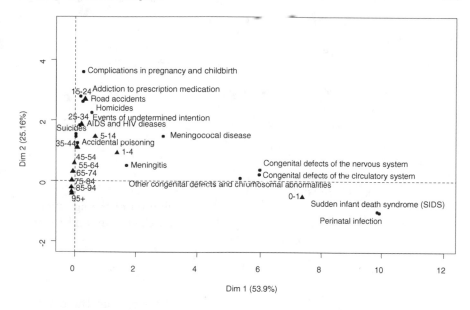

FIGURE 2.20

Mortality data: representation of age groups and the most specific causes of mortality on the first plane.

tions for the rows and columns for the different dimensions. They are expressed as percentages (and are therefore sometimes referred to as relative contributions). We present those in the columns in their "natural" order, thus:

```
> round(res.ca$col$contrib[,1],3)
   0-1    1-4    5-14   15-24  25-34  35-44
97.071  0.730  0.256  0.240  0.122  0.024
 45-54  55-64  65-74  75-84  85-94   95+
 0.004  0.068  0.306  0.660  0.451  0.069
```

The contributions confirm that the age group 0–1 year contributes (almost) entirely to the first dimension (as suggested by Figure 2.19); at this age, therefore, causes of mortality are extremely specific. This result supports that relating to the inertia of this age group in the overall space (0.5262) commented above.

As there are many causes of death, we present their contributions in the descending order, limiting ourselves to the five causes with the greatest contributions (these five causes account for 95.56% of the construction of the first dimension). Thus:

```
> res.ca$row$contrib[rev(order(res.ca$row$contrib[,1])),1]
                         Perinatal infection  59.101
```

```
                                                   SIDS  14.440
                   Congenital defects of the circulatory system  11.512
Other congenital defects and chromosomal abnormalities   7.428
                 Congenital defects of the nervous system   3.079
```

The dimension highlights the specific causes of death (almost by definition, as seen in the terms "perinatal" and "infant" in the age group of 0 to 1 year. These contributions complete the graph in indicating the key role of infection.

2.10.5 Plane 2-3

The first dimension highlights the most prominent trait of the deviation from independence: causes of mortality specific to newborn babies. At this stage, two options are available:

1. As the specificity of this age range has been well established, this age group is removed from the analysis and the analysis conducted a second time. In doing so, the range of our study is altered: we focus specifically on the population of individuals of more than one year old. One is often tempted by this method, which breaks a wide-ranging domain down into simple elements prior to studying them.

2. Continue the investigation of this CA. The orthogonality of the dimension ensures that as the uniqueness of the 0 to 1 age group has been expressed on the first dimension, it will not go on to "distort" the following axes. This is the tactic that we chose to follow (and which, generally, we would recommend).

We shall now go on to consider plane 2-3 (see Figures 2.21 and 2.22). The representation of the age groups identifies a Guttman effect (or horseshoe effect). Such an effect may appear in the case of ordered categorical variables when one axis confronts the smallest categories to the highest and another axis confronts those extreme categories to the ones in between. The second dimension (abscissa axis) confronts the youngest age groups with the eldest age groups whereas the third dimension confronts the extreme ages with the average age groups.

Along dimension 2, the "adult" age groups ($\geq$ 15 years) are arranged according to their "natural" order. This indicates a regular increase of the mortality profile with age. This representation requires two further remarks:

1. The difference between the two curves of the parabola stem from the differences in sample sizes; the youngest categories are generally the rarest (and of course, in this case, this is not a criticism, as we are talking about the number of deaths), and the origin of the dimensions is located in the clouds' centre of gravity (of both rows and columns). It is effectively found near to the categories with the greatest sample size (the average life expectancy is 70.98 years,

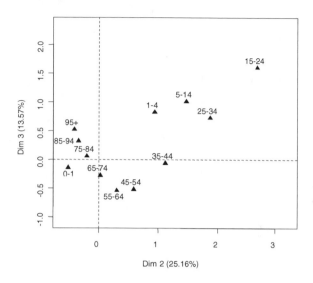

FIGURE 2.21
Mortality data: representation of age groups on plane 2-3.

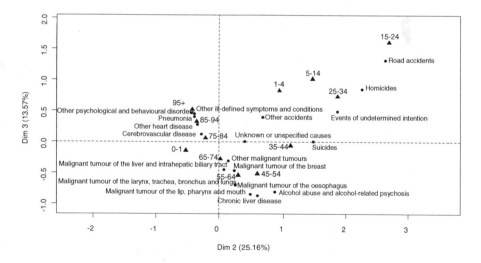

FIGURE 2.22
Mortality data: plane 2-3 with the representation of age groups and causes of mortality with a contribution greater than 1.5% on one of the two dimensions.

which corresponds to the results in the graph). Another approach to this difference in sample size between young and elderly age groups is that the latter are "automatically" closer to the average profile as they influence it the most.

2. Furthermore, the graph clearly shows that the upper age groups are closer to one another than those of younger adults.

We can validate this observation in the overall space by calculating the distances between age groups in this space. The command used to obtain this matrix, along with the matrix itself, is given below.

```
> res.ca<-CA(death,row.sup=c(66:nrow(death)),ncp=Inf)
> round(dist(res.ca$col$coord),3)
      0-1    1-4   5-14 15-24 25-34 35-44 45-54 55-64 65-74 75-84 85-94
1-4   6.818
5-14  7.221 2.069
15-24 7.965 3.656 2.008
25-34 7.611 3.263 1.874 1.840
35-44 7.495 3.241 2.118 2.694 1.250
45-54 7.480 3.322 2.352 3.166 1.944 0.874
55-64 7.483 3.354 2.428 3.329 2.171 1.175 0.412
65-74 7.480 3.346 2.428 3.374 2.249 1.343 0.767 0.445
75-84 7.480 3.342 2.445 3.410 2.312 1.496 1.073 0.827 0.422
85-94 7.486 3.351 2.485 3.449 2.373 1.619 1.282 1.094 0.754 0.380
95+   7.505 3.390 2.562 3.508 2.463 1.766 1.491 1.355 1.098 0.807 0.474
```

First, this matrix illustrates the great distance between the 0 to 1 age group and the others, as shown on the first dimension. Furthermore, it indicates that the distance between consecutive age groups decreases steadily between 1 year and 54 years, after which it stabilises around a weak value. This is in line with our observation on plane 2-3 concerning the age groups above 15 years (for 1 to 4 and 5 to 15 years, other dimensions, including the first, are required to account for this specificity).

The contributions to the constructions of dimensions and the representation qualities are as follows for the different age groups:

```
> round(cbind(res.ca$col$contrib[,2:5],res.ca$col$cos2[,2:5]),3)
```

	Contributions				Representation Quality (cos2)			
	Dim 2	Dim 3	Dim 4	Dim 5	Dim 2	Dim 3	Dim 4	Dim 5
0-1	1.060	0.146	0.015	0.599	0.005	0.000	0.000	0.000
1-4	0.711	1.031	2.089	58.057	0.083	0.065	0.032	0.523
5-14	2.659	2.375	4.075	15.458	0.401	0.193	0.081	0.180
15-24	33.216	21.793	13.518	0.920	0.684	0.242	0.037	0.001
25-34	18.946	5.357	4.207	6.381	0.771	0.118	0.023	0.020
35-44	12.049	0.074	19.113	1.596	0.759	0.003	0.159	0.008
45-54	9.017	12.762	11.460	2.453	0.498	0.380	0.083	0.010
55-64	3.585	20.883	0.002	2.923	0.227	0.713	0.000	0.014
65-74	0.038	10.562	11.896	0.471	0.005	0.690	0.190	0.004
75-84	5.439	0.719	9.790	5.097	0.573	0.041	0.136	0.042
85-94	10.447	16.309	6.272	0.298	0.492	0.414	0.039	0.001
95+	2.832	7.988	17.564	5.747	0.253	0.385	0.207	0.040

For causes of death, contributions are sorted in descending order and the five strongest contributions are given for dimensions 2 and 3.

```
> cbind(res.ca$row$contrib[,2], res.ca$row$cos2[,2],res.ca$call$marge.row)
  [rev(order(res.ca$row$contrib[,2])),]
```

	contrib	cos2	Sample size in %
Road accidents	41.048	0.754	0.015
Suicides	16.250	0.818	0.019
Other cardiomyopathies	4.272	0.546	0.092
Other accidents	4.130	0.592	0.024
Events of undetermined intention	3.390	0.886	0.003

```
> cbind(res.ca$row$contrib[,3],res.ca$row$cos2[,3],res.ca$call$marge.row)
  [rev(order(res.ca$row$contrib[,3])),]
```

	contrib	cos2	Sample size in %
Road accidents	19.199	0.190	0.015
Malignant tumour of the larynx, trachea, ...	16.503	0.818	0.048
Chronic liver disease	12.206	0.625	0.022
Other ill-defined symptoms and conditions	5.312	0.351	0.036
Other cardiomyopathies	5.071	0.349	0.092

Along the second dimension, the age groups between 15 and 44 years account for a combined contribution of 64.211%, and is therefore the basis for the interpretation of the results. The contributions of these three age groups therefore support the coordinates (the three sample sizes are similar). The age group 15–24 is the most extreme example that we should focus our attention in order to illustrate the dimension.

Road accidents contribute the most to this dimension (41.05%) and have the coordinate with the highest value. This cause of death is characteristic of young adults (high-value coordinate). This, along with its relatively high frequency (see Figure 2.17) means that young adults account for an essential dimension (the second of the deviation from independence (high contribution). This can be directly illustrated using the data (see Table 2.12): the percentage of young people in the deaths caused by road accidents is much higher than the percentage of young people among the overall deaths.

A similar parallel can be drawn with "homicides," which also has a high coordinate value indicating that it is characteristic of young adults. However, the low frequency of this latter cause (see Figure 2.17) leads to a weak contribution (1.86%): it is therefore not a typical cause of death in young adults. Again, Table 2.12 effectively illustrates the results directly from the data. In comparison with "road accidents," the lowest percentage for the 15–25 age group for homicide (14.56% instead of 28.80%) fits the fact that "homicides" is positioned closer to the centre.

The cause "suicides" is notably less characteristic of young adults (a more central position along with the lower percentage among young adults than for the two previous causes); however, its relatively high frequency (1.93%) means that this cause makes a considerable contribution to the causes characterising the deaths of young adults.

TABLE 2.12
Mortality Data: Extract of Some of the Data Concerning the Causes
Characterising Death in Young Adults; Raw Data and Frequencies

	15–24	25–34	35–44	Other	Total
Road accidents	4 653	2451	1841	7211	16,156
Homicides	144	199	180	466	989
Suicides	1431	2693	3280	13,003	20,407
Other	6203	9415	21,299	983,288	1,020,205
	15–24	25–34	35–44	Other	Total
Road accidents	0.288	0.152	0.114	0.446	1.000
Homicides	0.146	0.201	0.182	0.471	1.000
Suicides	0.070	0.132	0.161	0.637	1.000
Other	0.006	0.009	0.021	0.964	1.000

2.10.6   Projecting the Supplementary Elements

There are many possible approaches to analysing the data from 1979 and
from 2006. Thus, we can conduct a CA for each of the two tables, or for
their combined table. In this event, we chose to introduce the annual tables
as supplementary rows in the combined CA. The advantage of this approach
is (1) that we do not conduct multiple analyses; (2) to simultaneously analyse
the two tables in an "average" framework which has already been interpreted.

Each supplementary row is associated with a couple (cause, year), which
we shall refer to as the "annual-cause."

Figure 2.23 illustrates the evolution of a few causes of mortality. A given
cause of death, corresponding to the combined years 1979 and 2006, is connected to the supplementary points for this same cause of death in 1979 and
in 2006. There is one unique property of CA when representing multiple profiles and their sum: the average point (i.e., that which corresponds to the
sum) is located at the barycentre of the points for which it is the sum, in this
case, the two points 1979 and 2006. Thus, for example, the point *addiction
to prescription medication 2006* is closer to the average point than the point
addiction to prescription medication 1979. There were therefore more deaths
attributed to "addiction to prescription medication" in 2009 (189) compared
to 1979 (33). In contrast, deaths from influenza have decreased sharply (117
in 2006 compared with 1062 in 1979).

Let us consider two annual causes relating to the same cause. Going
beyond their position in relation to the average point, it is above all their
distances on the plane which is informative, as they indicate an evolution of
the corresponding age profiles. The causes featured in Figure 2.23 were chosen
precisely because of the marked evolution between 1975 and 2006 in terms of
age profile. We shall briefly comment on two examples.

Addiction to Prescription Medication. The graph points to an evolution
of the age profile towards young people. This can be validated using the data
directly, by combining the ages into two groups to simplify matters: ≤ 44
years and > 44 years (the 44-year limit is suggested by the raw data). The
increase of this cause of death in young people is noticeable in terms of absolute

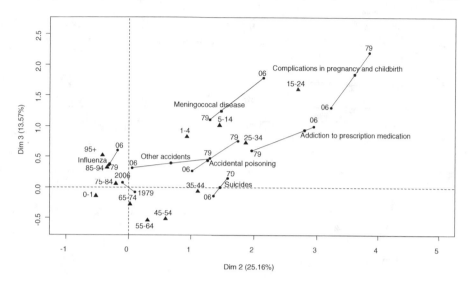

FIGURE 2.23
Mortality data: projection of a number of supplementary elements.

frequency (an increase from 13 to 167) or relative frequency (the percentage of this cause in young people increased from 39% to 88%, see Table 2.13).

TABLE 2.13
Mortality Data: Extract of Data Relation to Addiction to Medication

	Sample size			Percentage		
	15–44	Other	Total	15–44	Other	Total
79_Addiction to medication	13	20	33	0.394	0.606	1
06_Addiction to medication	167	22	189	0.394	0.606	1
Addiction to medication	180	42	222	0.811	0.189	1

As the sample sizes are rather small, it is best to validate the relationship between age and year using a χ^2 test conducted on the "Total" table at the bottom of Figure 2.15. This yields a value of 43.913 (with a p-value equal to 3.4×10^{-11}), which is highly significant.

Suicides. The graph suggests an evolution in the opposite direction to that described above, thus a slight decrease in this cause in young people. This evolution seems to be much less marked than that of the previous cause, however, as "suicide" has a high frequency; it nonetheless merits our attention. Table 2.14 confronts age (grouped into two categories, divided this time at 34 years, as suggested by the raw data) and year. It indicates that, between 1979 and 2006, the percentage of young people among those who commit suicide decreased from 24.6% to 16.0%. This evolution is less spectacular than that of addiction (the Φ^2 calculated from the table are 0.198 for the first year and 0.011 for the second); however, due to the larger sample size, they are even more significant (p-value less than 2.2×10^{-16}).

TABLE 2.14
Mortality Data: Extract of Data Relating to Suicides

	Sample size			Percentage		
	15–34	Other	Total	15–34	Other	Total
79_Suicides	2461	7531	9992	0.246	0.754	1.000
06_Suicides	1663	8752	10,415	0.160	0.840	1.000
Suicides	4124	16,283	20,407	0.202	0.798	1.000

Besides the "annual-causes," the average age profile (i.e., all causes of death combined) of each year can be introduced as supplementary. For the years 1979 and 2006, these profiles are the row profiles of the 1979 and 2006 tables. They are used to study the evolution of the distribution of deaths by age group in the period between the two years. Figure 2.23 shows that, between 1979 and 2006, the average age profile has moved towards higher age groups. This is due to (1) the ageing population (it must not be forgotten that the data are sample sizes and not rates); and (2) increasing life expectancy.

We have already shown that these data are actually available for every year between 1979 and 2006, whereas only these two years were introduced in the analysis so as not to complicate the results. However, it is possible, without complicating the analysis, to introduce the row margins for each yearly table as supplementary rows. The graph in Figure 2.24 is obtained by repeating the analysis with the same active elements but by introducing only the yearly age profiles as supplementary.

The sequence over the years is surprisingly regular, showing a steady evolution towards older age profiles, to such an extent that the irregularities in this trajectory merit our attention. For example, in 1999, the annual trajectory changes direction. Although the position of a row (i.e., a year) on the plane should be interpreted in terms of all of the columns (age groups), Figure 2.24 illustrates a trajectory which, until 1999, moves away from the age groups 45–54 and 55–64 years, and then no longer continues in this direction. An investigation of the evolution of deaths in the 45–64 age group (see Figure 2.25) does indeed show a decrease until 1999–2000, and then a slight increase after this date. This increase could well be due to the arrival in this age group of a generation of baby boomers.

Technically, there are two ways of constructing the graph in Figure 2.24:

1. We discard the supplementary elements which do not correspond to the total number of deaths per year between 1976 and 2006:

```
> res.ca$row.sup$coord <- res.ca$row.sup$coord[130:157,]
> plot.CA(res.ca,invisible=c("row","col"),axes=2:3)
> points(res.ca$row.sup$coord[,2:3],type="l")
```

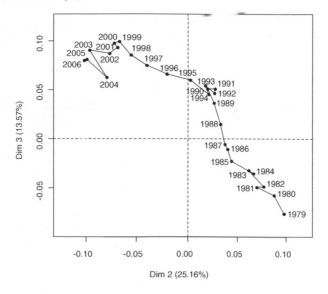

FIGURE 2.24
Mortality data: evolution of the total number of deaths by year and by age group.

2. We conduct another CA with only the total number of deaths per year between 1976 and 2006 as supplementary rows. We then construct another graph without the active elements using the argument `invisible=c("row","col")`. Thus, we visualise the supplementary elements alone, and then connect the points:

```
> tab.evol <- death[-(66:194),]
> res.evol <- CA(tab.evol,row.sup=66:nrow(tab.evol),graph=FALSE)
> plot.CA(res.evol,invisible=c("row","col"),axes=2:3)
> points(res.evol$row.sup$coord[,2:3],type="l")
```

2.10.7 Conclusion

This example effectively illustrates the nature of the summaries that CA can yield from a complex table. The dimensions can highlight both specific cases if they present specific qualities (the 0–1 year age group) and more general phenomena.

The choice of active and supplementary elements is crucial and represents a precise objective. There are a number of different choices to be made. Within a period of learning new methodologies, or collecting data, the user can confront multiple approaches. When communicating the results, the user should then choose one of these approaches to avoid being overwhelmed. It

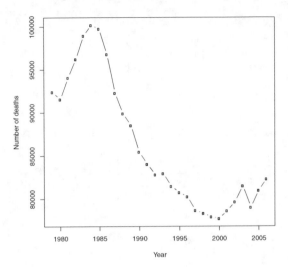

FIGURE 2.25
Mortality data: evolution of deaths in the 45–64 age group.

is thus vital to clearly specify the objective of the chosen analysis. In the example of the tables of annual deaths, let us compare chosen methodology (analysis of overall table and introduction of annual tables as supplementary), with a second methodology (analysis of a juxtaposition – in a column – of the annual tables and introduction of the overall table as supplementary).

As has already been noted, the CA of the overall table examines the relationship between the variables "cause" and "age" over the given period. The annual evolution of this relationship is then studied via that of the age profiles of the causes of death. In this analysis, the evolutions which do not fit into the overall relationship (i.e., over the whole period) cannot be illustrated.

The second approach, a CA of the juxtaposition in a column (see Figure 2.15), deals with the overall relationship and its evolution through that of the cause profiles. Presenting the objective this way is more pertinent than presenting it based on the relationship between age and the variable confronting cause and time, which is more formal although fundamental to CA. In this analysis, the specific dimensions of the annual evolution (i.e., nonrelated to the overall relationship) can be illustrated.

In addition, it must be noted that the annual table can also be brought together in rows, showing the evolution of the age-cause relationship through that of the death profiles for different age groups. This is suggested by a third CA (the annual tables brought together in an active row), but also by introducing annual age groups as supplementary columns in the CA of the overall table. This first analysis is again enhanced whilst at the same time retaining its simplicity (due to that of the active elements) and is recommended, at least during the first stages.

Finally, it must be noted that CA (like other multidimensional data analysis methods) simply provides a way of visualising the data. This visualisation is highly valuable and suggests interpretations that cannot be inferred from the raw data, without disproving them. The example presented here effectively illustrates this quality by highlighting the overall annual evolution, but CA tells us nothing about the evolution of the population pyramid or the evolution of mortality rates by cause or age range. CA helped us to answer the initial question (what is the relationship between age and causes of mortality?) but in the end raises yet more questions. Users may therefore remain hungry for further answers, but is this not symptomatic of all research?

3

Multiple Correspondence Analysis (MCA)

3.1 Data — Notation — Examples

Multiple Correspondence Analysis (MCA) is not really a new mathematical method but rather the specific application of correspondence analysis (CA) to tables with individuals and their answers to a number of categorical variables. It is nonetheless considered a separate method due to its unique properties and the interesting results obtained when using it. MCA is applied to tables with individuals in the rows and categorical variables in the columns. It is most commonly used to analyse data obtained through surveys: in such a context, each question corresponds to a variable and each possible answer to the question corresponds to a category of that variable. For example, eight possible answers (categories) are associated with the question "Which of these categories best describes your profession?" farmer, student, manual labourer, professional, senior management, employee, other profession, unemployed. For each of these variables, the individual must choose one single category.

In the following example section (Section 3.9), we will detail a few applications of MCA for data which does not come from surveys.

We denote x_{ij} the category chosen by the individual i for variable j; i varies from 1 to I and j from 1 to J. We consider categorical variable j to have K_j categories.

This chapter will be illustrated using a survey of 300 tea drinkers. The battery of questions with which they were presented dealt with their tea-drinking behaviour, the image they have of the product, and finally a few descriptive questions. In the following analysis, tea-drinking behaviours are the only active variables, whereas image and descriptive data are supplementary.

There are 19 questions relating to tea-drinking behaviour.

1. "What kind of tea do you drink the most (black tea, green tea, flavoured tea)?"

2. "How do you take your tea (nothing added, with lemon, with milk, other)?"

3. "What kind of tea do you buy (tea bags, loose tea, both)?"

4. "Do you add sugar to your tea (yes, no)?"

5. "Where do you buy your tea (in the supermarket, in specialist shops, both)?"

6. "What kind of tea do you buy (cheapest, supermarket brand, well-known brand, luxury, it varies, I don't know)?"

7. "How often do you drink tea (more than twice a day, once a day, 3 to 6 times a week, once or twice per week)?"

8–13. Six questions deal with the location in which the tea is drunk: "Do you drink tea at home?" "Do you drink tea at work?" "Do you drink tea in tearooms or coffee shops?" "Do you drink tea at friends' houses?" "Do you drink tea in restaurants?" "Do you drink tea in bars?" For each of these six questions, the participants were required to answer *yes* or *no*.

14–19. Six questions deal with the time of day when the tea is drunk: "Do you drink tea at breakfast?" "Do you drink tea in the afternoon?" "Do you drink tea in the evening?" "Do you drink tea after lunch?" "Do you drink tea after dinner?" "Do you drink tea throughout the day?" For each of these six questions, the participants were required to answer *yes* or *no*.

Twelve questions were asked about the image of the product: "Do you consider tea to be exotic?" "Do you associate tea with spirituality?" "Is tea good for your health?" "Is tea a diuretic?" "Do you associate tea with friendliness?" "Does tea stop the body from absorbing iron?" "Is tea feminine?" "Is tea refined?" "Will tea help you to lose weight?" "Is tea a stimulant?" "Is tea a relaxant?" "Does tea have any effect on your overall health?" For each of these 12 questions, the participants were required to answer *yes* or *no*.

Four descriptive questions were also asked: sex, professional category (farmer, manual labourer, professional, senior management, employee, other profession, student, unemployed), age, and whether or not the participant regularly played sports (*yes* or *no*).

3.2 Objectives

The data may be studied according to the individuals, the variables, and the categories. This raises a certain number of questions relating to each of these very different aspects.

3.2.1 Studying Individuals

Studying the individuals means understanding the similarities between the individuals in terms of all the variables. In other words, to provide a typology of the individuals: which are the most similar (and most dissimilar) individuals? Are there groups of individuals which are consistent in terms of their similarities? In the example, two tea drinkers are considered similar if they answered the questions in the same way.

Individuals are compared on a basis of presence–absence of the categories which they selected. From this perspective alone, the distance between two individuals depends entirely on their characteristics and not on those of the other individuals. However, it is important to account for the characteristics of the other individuals when calculating this distance.

Let us consider four examples in order to understand how the distance between two individuals might be calculated:

1. If two individuals select the same categories, the distance which separates should be nil.

2. If two individuals both select a lot of the same categories, they should be close together.

3. If two individuals select all of the same categories except for one which is selected by one of the individuals and only rarely by all of the other individuals, they should be distanced to account for the uniqueness of one of the two.

4. If two individuals share a rare category, they should be close together despite their differences elsewhere in order to account for their common distinctiveness.

These different examples can be used to show that the individuals must be compared category by category whilst at the same time taking into account the rarity or the universal nature of that category.

3.2.2 Studying the Variables and Categories

As in principal component analysis (PCA), the aim is to summarise the relationships between the variables. These relationships are either studied in pairs (see Chapter 2 on CA) or all together. In this last case, we are looking for synthetic variables which sum up the information contained within a number of variables. The information carried by a variable can be studied in terms of its categories. In MCA, we focus primarily on studying the categories, as categories represent both variables and a group of individuals (all of the individuals who select this category).

To study how close the categories are to one another, one must first determine the distance between the categories. Thus, two categories k and k' are each associated with a group of individuals. One way of comparing these two categories is to count the individuals which select both categories: two categories can therefore be said to be further away (in terms of distance) the fewer individuals they have in common. In other words, the number of individuals which has either category k or category k' is rather high; this number is expressed $I_{k \neq k'}$.

However, it is important to account for the size of each group of individuals when calculating this distance. Let us consider an example with three categories k, k', and k'', each composed of 10, 100, and 100 individuals, respectively. If categories k and k' share no individuals, $I_{k \neq k'} = 110$. If categories k and k' share 45 individuals, $I_{k' \neq k''} = 110$. However, assume that k and k' share 0% individuals whereas k' and k'' share 45% individuals. Categories k and k' need to be further apart than categories k' and k''. It is therefore important to account for the sample size for each category.

3.3 Defining Distances between Individuals and Distances between Categories

As seen above in Section 2.2 on Objectives, we focus mainly on individuals and categories when analysing a table of individuals × categorical variables. From a data table of individuals × variables it is therefore natural to construct an indicator (dummy) matrix with individuals in the rows and all of the categories for every variable in the columns. The element x_{ik} of this table has a value of 1 if individual i carries category k, and 0 if it does not. This table has $I \times K$ dimensions (with $K = \sum_{j=1}^{J} K_j$) and is composed entirely of 0 and 1.

3.3.1 Distances between the Individuals

By using an indicator matrix along with the objectives outlined above, the distances between individuals can be calculated by adding the differences between categories, $(x_{ik} - x_{i'k})^2$, and counterbalanced using a function inversely proportional to I_k (with I_k the number of individuals carrying category k). This distance (squared) is expressed as

$$d_{i,i'}^2 = C \sum_{k=1}^{K} \frac{(x_{ik} - x_{i'k})^2}{I_k},$$

where C is a constant.

3.3.2 Distances between the Categories

The distance between two categories k and k' is calculated by counting the individuals which carry either category k or category k' (such as $I_{k \neq k'}$), and counterbalancing using a function inversely proportional to I_k and $I_{k'}$. This distance can therefore be expressed as

$$d^2_{k,k'} = C' \frac{I_{k \neq k'}}{I_k I_{k'}},$$

where C' is a constant. However, according to the encoding ($x_{ik} = 0$ or 1), the number of individuals carrying only one of the two categories is equal to $I_{k \neq k'} = \sum_{i=1}^{I}(x_{ik} - x_{ik'})^2$. Thus

$$d^2_{k,k'} = C' \frac{1}{I_k I_{k'}} \sum_{i=1}^{I}(x_{ik} - x_{ik'})^2.$$

In further developing this equation, we obtain

$$
\begin{aligned}
d^2_{k,k'} &= C' \frac{1}{I_k I_{k'}} \sum_{i=1}^{I}(x^2_{ik} + x^2_{ik'} - 2x_{ik}x_{ik'}), \\
&= C' \frac{\sum_{i=1}^{I} x^2_{ik} + \sum_{i=1}^{I} x^2_{ik'} - 2\sum_{i=1}^{I} x_{ik}x_{ik'}}{I_k I_{k'}}.
\end{aligned}
$$

By using the encoding properties ($x_{ik} = 0$ or 1 and therefore $x^2_{ik} = x_{ik}$ and thus $\sum_i x^2_{ik} = \sum_i x_{ik} = I_k$), we obtain

$$d^2_{k,k'} = C' \left(\frac{1}{I_{k'}} + \frac{1}{I_k} - 2\frac{\sum_{i=1}^{I} x_{ik}x_{ik'}}{I_k I_{k'}} \right).$$

However,

$$\frac{1}{I_k} = \frac{I_k}{I^2_k} = \frac{\sum_{i=1}^{I} x^2_{ik}}{I^2_k}.$$

The distance (squared) between two categories can therefore be expressed as

$$
\begin{aligned}
d^2_{k,k'} &= C' \left(\frac{\sum_{i=1}^{I} x^2_{ik'}}{I^2_{k'}} + \frac{\sum_{i=1}^{I} x^2_{ik}}{I^2_k} - 2\frac{\sum_{i=1}^{I} x_{ik}x_{ik'}}{I_k I_{k'}} \right), \\
&= C' \left(\sum_{i=1}^{I} \left(\frac{x_{ik'}}{I_{k'}} \right)^2 + \sum_{i=1}^{I} \left(\frac{x_{ik}}{I_k} \right)^2 - 2\sum_{i=1}^{I} \left(\frac{x_{ik}}{I_k} \times \frac{x_{ik'}}{I_{k'}} \right) \right), \\
&= C' \sum_{i=1}^{I} \left(\frac{x_{ik}}{I_k} - \frac{x_{ik'}}{I_{k'}} \right)^2.
\end{aligned}
$$

3.4   CA on the Indicator Matrix

3.4.1   Relationship between MCA and CA

If, in the following expressions, we consider that the constant $C = I/J$, the distance (squared) between two individuals i and i' is expressed as

$$
\begin{aligned}
d_{i,i'}^2 &= \frac{I}{J} \sum_{k=1}^{K} \frac{1}{I_k} \left(x_{ik} - x_{i'k}\right)^2, \\
&= \sum_{k=1}^{K} \frac{IJ}{I_k} \left(\frac{x_{ik}}{J} - \frac{x_{i'k}}{J}\right)^2, \\
&= \sum_{k=1}^{K} \frac{1}{I_k/(IJ)} \left(\frac{x_{ik}/(IJ)}{1/I} - \frac{x_{i'k}/(IJ)}{1/I}\right)^2.
\end{aligned}
$$

When including the notations from the contingency table introduced in CA and applied to the indicator matrix, we obtain

$$
f_{ik} = x_{ik}/(IJ),
$$

$$
f_{\bullet k} = \sum_{i=1}^{I} x_{ik}/(IJ) = I_k/(IJ),
$$

$$
f_{i\bullet} = \sum_{k=1}^{K} x_{ik}/(IJ) = 1/I.
$$

Here we can identify the χ^2 distance between the row profiles i and i' calculated from the indicator matrix:

$$
d_{\chi^2}^2(\text{row profile } i, \text{row profile } i') = \sum_{k=1}^{K} \frac{1}{f_{\bullet k}} \left(\frac{f_{ik}}{f_{i\bullet}} - \frac{f_{i'k}}{f_{i'\bullet}}\right)^2.
$$

Furthermore, if we assume that the constant $C' = I$, the distance (squared) between two categories k and k' is expressed as

$$
\begin{aligned}
d_{k,k'}^2 &= I \sum_{i=1}^{I} \left(\frac{x_{ik}}{I_k} - \frac{x_{ik'}}{I_{k'}}\right)^2, \\
&= \sum_{i=1}^{I} \frac{1}{1/I} \left(\frac{x_{ik}/(IJ)}{I_k/(IJ)} - \frac{x_{ik'}/(IJ)}{I_{k'}/(IJ)}\right)^2.
\end{aligned}
$$

Here we can identify the distance of χ^2 between the column profiles k and k' calculated from the indicator matrix:

$$d^2_{\chi^2}(\text{column profile } k, \text{column profile } k') = \sum_{i=1}^{I} \frac{1}{f_{i\bullet}} \left(\frac{f_{ik}}{f_{\bullet k}} - \frac{f_{ik'}}{f_{\bullet k'}} \right)^2.$$

The "cautious" choice of constants C and C' stems from the χ^2 distances of the row and column profiles, thus relating back to correspondence analysis. In terms of calculations (i.e., by the programme), MCA is therefore based on a correspondence analysis applied to an indicator matrix.

3.4.2 The Cloud of Individuals

Once the cloud of individuals has been constructed, as in CA (creation of profiles, distance from χ^2, weight = margin), it is represented using the same principal component approach as seen previously for PCA and CA: maximising the inertia of the cloud of individuals projected onto a series of orthogonal axes (see Section 3.6 on implementation).

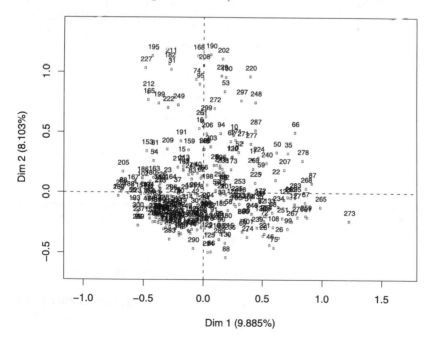

FIGURE 3.1
Tea data: plane representation of the cloud of individuals.

The graph of individuals for the first two principal components (17.99% of the explained inertia) is provided for the tea example in Figure 3.1. As in most analyses of survey data, the cloud of individuals is made up of many points and our objective is to see if we can identify a specific shape, or groups of notable individuals. In the example, there are no notable groups of individuals: the cloud of points is a rather consistent shape.

To illustrate the notion of distance between individuals, we shall consider the following four individuals: 200, 262 (at the negative extremity of the first principal component) and 265, 273 (at the positive extremity of the first principal component). Individuals 200 and 262 (265 and 273, respectively) are close as they share a lot of common categories. Pairs 200–262 and 265–273 are far from one another (opposed on the first axis) as they share only very few categories (see Figure 3.2).

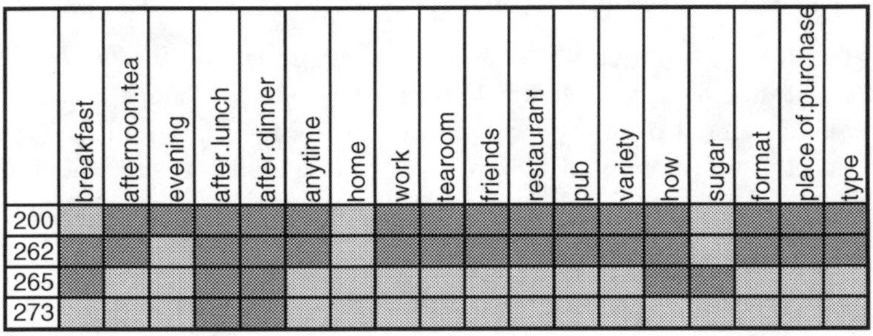

	breakfast	afternoon.tea	evening	after.lunch	after.dinner	anytime	home	work	tearoom	friends	restaurant	pub	variety	how	sugar	format	place.of.purchase	type
200																		
262																		
265																		
273																		

FIGURE 3.2
Tea data: comparison of individuals 200, 262, 265, and 273 (light gray = presence of the category).

As in any principal component method, the dimensions of the MCA can be interpreted using the individuals. Individuals 265 and 273 both drink tea regularly and at all times of day. Individuals 200 and 262 only drink tea at home, either at breakfast or during the evening. This exploratory approach can be tedious due to the large number of individuals, and is generalised by studying the categories through the individuals that they represent.

3.4.3 The Cloud of Variables

The variables can be represented by calculating the correlation ratios between the individuals' coordinates on one component and each of the categorical variables. If the correlation ratio between variable j and component s is close to 1, the individuals carrying the same category (for this categorical variable) have similar coordinates for component s. The graph of variables for the tea example is shown in Figure 3.3.

The variables *type*, *format*, and *place of purchase* are closely related to each of the first two components, although it is unclear how (this appears in

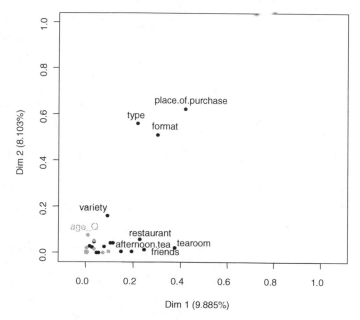

FIGURE 3.3
Tea data: plane representation of the cloud of variables.

the representation of the categories). This graph is especially useful to clear the way when faced with a large number of variables.

3.4.4 The Cloud of Categories

As with the supplementary categorical variables in PCA, the categories can be represented at the barycentre of the individuals with those categories. This representation is optimised as it corresponds to the representation obtained by maximising the inertia of the cloud of categories on a sequence of orthogonal axes by homothetic transformations (see Section 3.4.5).

The graph of categories for the tea example is shown in Figure 3.4. The first component opposes the categories *tearoom, supermarket + specialist shop, teabag + loose, bar, restaurant, work* with the categories *not.with.friends, not.work, not.restaurant, not.home*. This first component thus opposes regular tea drinkers with those who drink tea only occasionally. The second component distinguishes the categories *specialist shop, loose, luxury* and, to a lesser extent, *green* and *after dinner*, from all of the other categories.

Remark
The barycentre of all of the categories for a given variable is located at the cloud of individuals' centre of gravity. It therefore merges with the origin of the axes.

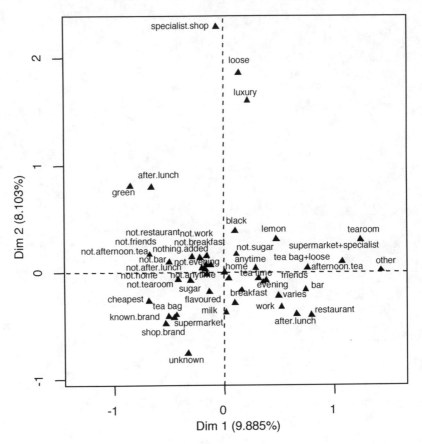

FIGURE 3.4
Tea data: plane representation of the cloud of categories.

The inertia for a category k can be calculated from the distance (squared) from k to the centre of gravity of the cloud of categories with coordinates of $1/I$ (i.e., the average vector of all of the categories):

$$
\begin{aligned}
d_{k,G_K}^2 &= I \sum_{i=1}^{I} \left(\frac{x_{ik}}{I_k} - \frac{1}{I} \right)^2, \\
&= I \left(\sum_{i=1}^{I} \frac{x_{ik}^2}{I_k^2} - \frac{2}{I} \frac{x_{ik}}{I_k} + \frac{I}{I^2} \right), \\
&= I \left(\frac{1}{I_k} - \frac{2}{I} + \frac{1}{I} \right),
\end{aligned}
$$

$$= \frac{I}{I_k} - 1.$$

This distance increases as only very few individuals have taken the category k.

It must be noted that, as in CA, the weight of a column profile corresponds to its margin (here, $I_k/(IJ)$). The inertia of category k can therefore be expressed as

$$\text{Inertia}(k) = d_{k,G_K}^2 \times \frac{I_k}{IJ} = \frac{I_k}{IJ}\left(\frac{I}{I_k} - 1\right) = \frac{I - I_k}{IJ} = \frac{1}{J}\left(1 - \frac{I_k}{I}\right).$$

This formula shows that the inertia of a category increases when the category is rare: for example, if 1% of individuals carry category k and 50% carry category k', the inertia associated with k will be approximately twice that of k'. It is therefore common for the principal dimensions of the MCA to be generated by a few rare categories. This is almost systematic if these rare categories are shared by the same individuals, which is quite common when these categories are missing data (for example, an individual who did not reply to several questions in a survey). As the components are constructed according to only a few individuals, in such cases it may be preferable to eliminate these rare categories in order to focus on the phenomena at hand. To do so, it is possible to group together certain categories, which is quite natural, particularly in the case of sequential categories (for example, it would be possible to group *60–75 years* along with *over 75 years*). It is also possible to randomly distribute the individuals associated with rare categories within other categories (whilst respecting the proportions associated with each one). This method is known in France as "ventilation" (see Section 3.7.1.2).

The inertia of all of the K_j categories for a variable j, known as the inertia of variable j, is equal to

$$\text{Inertia}(j) = \sum_{k=1}^{K_j} \frac{1}{J}\left(1 - \frac{I_k}{I}\right).$$

As $\sum_{k=1}^{K_j} I_k = I$, we obtain

$$\text{Inertia}(j) = \frac{K_j - 1}{J}.$$

Thus, the inertia of a variable only depends on the number of categories that make it up: it is therefore greater when this number is higher. In the example, the variable *type* (which has six categories) has five times more inertia than the variable *sugar* (which has two categories).

Remark

It is recommended that the questionnaires be constructed with an equal number of answers to each question (in order to have an equal number of categories for each variable) but this is merely a recommendation and not a requirement.

Indeed, in practice, if a variable has a large number of categories, these categories are shared over many different dimensions (number of dimensions = number of categories minus 1). Therefore, this variable will not systematically influence the construction of the principal components.

Finally, the inertia associated with all of the categories can be calculated, and also corresponds to the inertia of the cloud of categories (N_K):

$$\text{Inertia}(N_K) = \sum_{j=1}^{J} \frac{K_j - 1}{J} = \frac{K}{J} - 1.$$

This inertia only depends on the structure of the questionnaire, or more precisely, the average number of categories per variable. For example, if all of the variables have the same number of categories ($\forall j, \ K_j = c$), the inertia of the cloud will be equal to $c - 1$.

3.4.5 Transition Relations

As in PCA or CA, transition relations link the cloud of individuals N_I to the cloud of categories N_K. In the following transition formulae, obtained by applying CA relations to an indicator matrix, $F_s(i)$ ($G_s(k)$, respectively) designates the coordinate for individual i (and category k, respectively) on the component of rank s.

$$F_s(i) \quad = \quad \frac{1}{\sqrt{\lambda_s}} \sum_{j=1}^{J} \sum_{k=1}^{K_j} \frac{x_{ik}}{J} G_s(k),$$

$$G_s(k) \quad = \quad \frac{1}{\sqrt{\lambda_s}} \sum_{i=1}^{I} \frac{x_{ik}}{I_k} F_s(i).$$

On the component of rank s, up to the multiplicative factor $\frac{1}{\sqrt{\lambda_s}}$, the first relationship expresses that individual i is at the centre of gravity of the categories that it carries (as $x_{ik} = 0$ for the categories that it does not carry).

On the component of rank s, up to the multiplicative factor $\frac{1}{\sqrt{\lambda_s}}$, the second relationship expresses that category k is at the centre of gravity of the individuals that carry it.

As the categories correspond to groups of individuals, it seems logical to represent them on the graph of individuals. The transition relations show that two representations are possible: to draw the categories at the centre of gravity of the individuals, or to draw the individuals at the centre of gravity of the categories. Both of these graphs are interesting but, as in CA, it is impossible to obtain both of these properties simultaneously. To compromise, we therefore construct a graph as follows: the graph of individuals is constructed, and we then position the categories by multiplying their coordinates on the component of rank s by the $\sqrt{\lambda_s}$ coefficient (see Figure 3.5).

As a result, the cloud of categories is reduced by a different coefficient for each component. This graph means we can avoid having all of the categories concentrated around the centre of the diagram. It should, however, be noted that most often we refer quickly to the shape of the cloud of individuals (most of the time the individuals are anonymous) prior to interpreting the cloud of categories in detail.

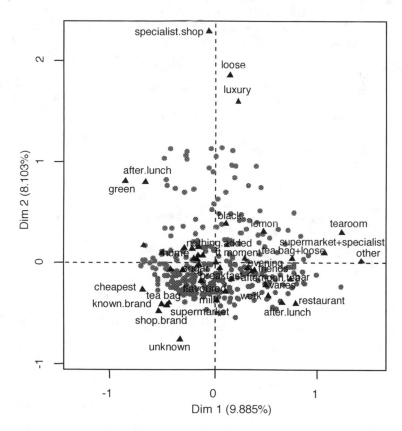

FIGURE 3.5
Tea data: plane representation of the cloud of individuals (grey points) and categories.

The second transition relation meets the objective fixed in Section 3.2.2: two categories are similar if they are carried by the same individuals. It also suggests one way of interpreting this proximity between two categories when they belong to the same variable. In this case, both categories cannot be chosen by one individual (limited choice) thus distancing them from one another by design. However, as each category represents a group of individuals, two groups of individuals can be close if they have the same profiles elsewhere.

In the example, the categories *supermarket brand* and *well-known brand* associated with the question "What type of tea do you buy (cheapest, supermarket brand, well-known brand, luxury, it varies, I don't know)?" are represented side by side (see Figure 3.4). In fact, these two categories group together consumers with similar profiles: they both tend to buy tea in supermarkets rather than in specialist shops, to drink tea in teabags, and to add sugar to their tea (see Table 3.1). The influence of all of these variables brings these two categories closer together, and this multidimensional approach prevails over the exclusive nature of the responses to each question.

TABLE 3.1
Tea Data: Comparison of Consumers Buying Well-Known Brands (and Supermarket Brands, Respectively) with the Average Profile

	Well-known brand	Supermarket brand	Overall
Place of purchase=supermarket	86.32%	95.24%	64.00%
Format=tea bag	73.68%	76.19%	56.67%
Sugar=yes	52.63%	61.90%	48.33%
Format=tea bag+loose	21.05%	19.05%	31.33%
Place of purchase=specialist shop	2.11%	0.00%	10.00%
Place of purchase=supermarket+specialist	11.58%	4.76%	26.00%

Note: 86.32% (95.24%, respectively) of consumers who purchase well-known brands (and supermarket brands, respectively) buy tea in supermarkets compared with 64% for all participants.

3.5 Interpreting the Data

3.5.1 Numerical Indicators

3.5.1.1 Percentage of Inertia Associated with a Component

The percentage of inertia associated with a component is calculated in the same way as for any principal component method (see Section 1.6.1.1). In MCA, the percentages of inertia associated with the first components are generally much lower than in PCA. This is because, in PCA, only the linear relationships are studied: one single component should be sufficient to represent all of the variables if they are highly correlated. In MCA, we are studying much more general relationships and at least $\min(K_j, K_l) - 1$ dimensions are required in order to represent the relationship between two variables, each of which has K_j and K_l categories, respectively. It is therefore common for many more dimensions to be studied in MCA than in PCA.

In the example, 17.99% of the data are represented by the first two components (9.88% + 8.10% = 17.99%). It can be seen (Table 3.2 or Figure 3.6) that there is a steady decrease in eigenvalues. Here we shall analyse only the first two components, even if it might also be interesting to examine the next components.

TABLE 3.2
Tea Data: Decomposition of Variability for the First 10 Components

	Eigenvalue	Percentage of inertia	Total percentage of inertia
Dim 1	0.15	9.88	9.88
Dim 2	0.12	8.10	17.99
Dim 3	0.09	6.00	23.99
Dim 4	0.08	5.20	29.19
Dim 5	0.07	4.92	34.11
Dim 6	0.07	4.76	38.87
Dim 7	0.07	4.52	43.39
Dim 8	0.07	4.36	47.74
Dim 9	0.06	4.12	51.87
Dim 10	0.06	3.90	55.77

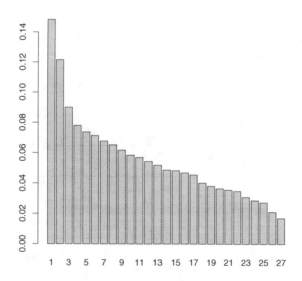

FIGURE 3.6
Tea data: chart of eigenvalues.

3.5.1.2 Contribution and Representation Quality of an Individual or Category

The calculations and interpretations for both contribution and representation quality for individuals and categories are the same as in CA. However, due to the scale of the dataset, representation quality on a given plane is often much weaker compared to the representation qualities obtained in CA (or PCA).

The larger scale of the dataset does not affect contribution, as this aspect is calculated for each component. It should be noted that the contribution of a categorical variable to a given component can be calculated by adding the contributions to these categories. The contribution (to the component of rank s) of a categorical variable divided by $J\lambda_s$ is equal to the correlation between the principal component and the categorical variable. In PCA, the vector of the coordinates for the individuals on the component of rank s is referred to as the principal component; this concept is transferred directly into MCA.

3.5.2   Supplementary Elements

As in PCA, supplementary elements may be individuals or categorical and/or quantitative variables.

For a supplementary individual i' and a supplementary category k', the transition formulae are expressed as

$$F_s(i') = \frac{1}{\sqrt{\lambda_s}} \sum_{j=1}^{J} \sum_{k=1}^{K_j} \frac{x_{i'k}}{J} G_s(k),$$

$$G_s(k') = \frac{1}{\sqrt{\lambda_s}} \sum_{i=1}^{I} \frac{x_{ik'}}{I_{k'}} F_s(i).$$

These transition formulae are identical to those of the active elements (individuals and categories). In the example (see Figure 3.7), the categories of the variables relating to the image of tea can be projected. These categories are at the centre of the graph, which shows that it will be rather difficult to connect, on the one hand, the behavioural variables with, on the other, the image and descriptive variables.

The quantitative supplementary variables are represented in the same way as in PCA (see Section 1.6.2): on a correlation circle using correlation coefficients between the variable and the principal components. In the example, the correlation circle (see Figure 3.8) is used to represent the quantitative variable *age*. This variable is not well represented; however, the correlation with the second principal component (0.204) is significant due to the great number of individuals. Young people are less likely to buy their tea in specialist shops. It can therefore also be said that older people are more likely to buy luxury loose tea in specialist shops.

Remark

The variable *age* was divided into categories (*15–24 years, 25–34 years, 35–44 years, 45–59 years, 60 years and over*) and is represented as a categorical supplementary variable. This division may help us to highlight nonlinear relationships. If we "zoom in" on the supplementary categories (see Figure 3.9), we can see that the categories of the variable *age* are arranged in their natural

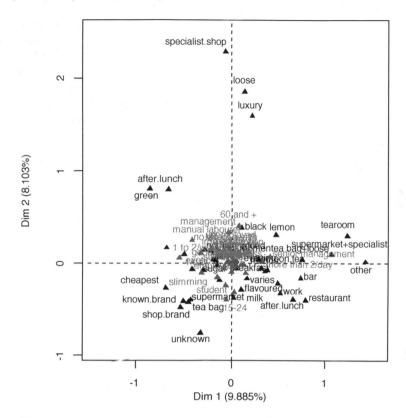

FIGURE 3.7
Tea data: representation of active and supplementary categories.

order along the second component (see Figure 3.9). This supports the positive correlation between the variable *age* and the second principal component.

3.5.3 Automatic Description of the Components

In the same way as in PCA (see Section 1.6.3), the components provided by MCA can be described automatically by all of the variables, be they quantitative or categorical (in this case we also use the categories), active or supplementary.

In the example (see Table 3.3), the first component is characterised by the variables *place of purchase*, *tearoom*, and so forth. In addition, we can see that certain supplementary variables are also related to this component (*sex* and *friendliness*). Since most variables have two categories, characterisation by category (see Table 3.4) is similar to that calculated from the variables, but

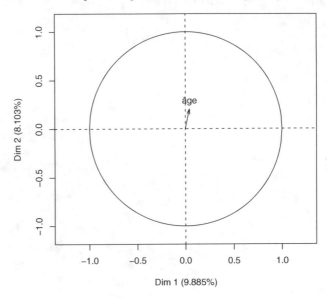

FIGURE 3.8
Tea data: representation of the supplementary variable *age*.

specifies the direction of the component: for example, the coordinate *tearoom* is positive whereas the coordinate *not tearoom* is negative. Individuals with more positive coordinates would therefore be more likely to go to tearooms.

TABLE 3.3
Tea Data: Description of the First Dimension by
Categorical Variables (Object `$'Dim 1'$quali`)

	R2	*p*-value
Place of purchase	0.4180	1.26e-35
Tearoom	0.3720	6.08e-32
Format	0.2990	1.27e-23
Friends	0.2430	8.62e-20
Restaurant	0.2260	2.32e-18
Afternoon tea	0.1920	1.65e-15
Type	0.2160	4.05e-14
Pub	0.1470	5.85e-12
Work	0.1120	3.00e-09
How	0.1030	4.80e-07
Variety	0.0895	8.97e-07
After lunch	0.0746	1.57e-06
Frequency	0.0944	1.85e-06
Friendliness	0.0713	2.71e-06
Evening	0.0531	5.59e-05
Anytime	0.0448	2.22e-04
Sex	0.0334	1.49e-03
After dinner	0.0329	1.61e-03
Breakfast	0.0254	5.67e-03
Sugar	0.0153	3.23e-02

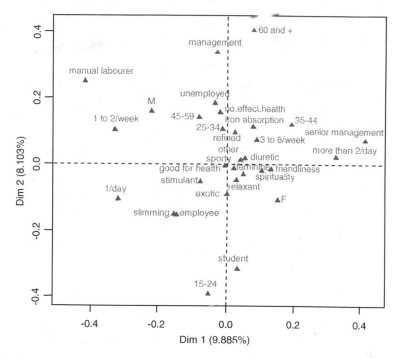

FIGURE 3.9
Tea data: representation of supplementary categories.

3.6 Implementation with FactoMineR

In this section, we shall illustrate how to conduct an MCA using FactoMineR and how to retrieve the results obtained from the tea dataset.

```
> library(FactoMineR)
> tea <- read.table("http://factominer.free.fr/bookV2/tea.csv",header=T,sep=";")
> summary(tea)
```

The MCA is obtained by specifying that, in this case, variable 22 is quantitative supplementary and variables 19 to 21 and 23 to 36 are categorical supplementary:

```
> res.mca<-MCA(tea,quanti.sup=22,quali.sup=c(19:21,23:36))
```

This command executes the MCA and produces the graph of variables (with both active and supplementary variables, see Figure 3.3), the graph of individuals (featuring the individuals and the categories for the active and supplementary variables, see Figure 3.5) as well as the graph of quantitative

TABLE 3.4
Tea Data: Description of the First Dimension by
Overexpressed Categories (Object `$'Dim 1'$category`)

	Estimate	*p*-value
Tearoom	0.2970	6.08e-32
Supermarket and specialist	0.3390	6.34e-23
Friends	0.2000	8.62e-20
Restaurant	0.2080	2.32e-18
Afternoon tea	0.1700	1.65e-15
Tea bag+loose	0.2350	2.72e-12
Pub	0.1810	5.85e-12
Work	0.1420	3.00e-09
Varies	0.2760	1.20e-07
After lunch	0.1490	1.57e-06
Friendliness	0.1300	2.71e-06
Other	0.3820	3.27e-06
Evening	0.0935	5.59e-05
Flavoured	0.1220	1.18e-04
More than 2/day	0.1490	1.30e-04
Anytime	0.0858	2.22e-04
Luxury	0.1710	7.32e-04
Black	0.1240	8.90e-04
Female	0.0716	1.49e-03
Not after dinner	0.1370	1.61e-03
Breakfast	0.0614	5.67e-03
Senior management	0.1680	6.09e-03
Not sugar	0.0476	3.23e-02

supplementary variables (see Figure 3.8). To construct a graph which features only certain specific elements, the function **plot.MCA** is used. The following commands are used to retrieve the graph of individuals (see Figure 3.1), along with those of active categories (see Figure 3.4), of superimposed representations (see Figure 3.5), of active and supplementary categories (see Figure 3.7), and of supplementary categories (see Figure 3.9):

```
> plot(res.mca,invisible=c("var","quali.sup"),cex=0.7)
> plot(res.mca,invisible=c("ind","quali.sup"))
> plot(res.mca,invisible="quali.sup")
> plot(res.mca,invisible="ind")
> plot(res.mca,invisible=c("ind","var"))
```

The table of eigenvalues (see Table 3.2) as well as the main results:

```
> summary(res.mca)
> lapply(dimdesc(res.mca),lapply,round,4)
```

The command **dimdesc** leads to an automatic description of the dimensions by the categorical variables (see Table 3.3) or the categories (see Table 3.4). The function **lapply** is only used to round data within a list (in this case within a list of lists!):

```
> lapply(dimdesc(res.mca),lapply,signif,3)
```

Additional Details. Confidence ellipses can be drawn around the categories of a categorical variable (i.e., around the barycentre of the individuals carrying that category), according to the same principle as that detailed for PCA (see Section 3.6). These ellipses are adapted to given plane representations and are used to visualise whether two categories are significantly different or not. It is possible to construct confidence ellipses for all of the categories for a number of categorical variables using the function **plotellipses** (see Figure 3.10):

```
> plotellipses(res.mca,keepvar=c("restaurant","place.of.purchase",
      "relaxant","profession"))
```

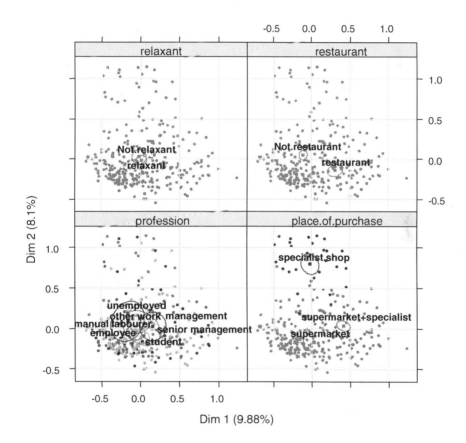

FIGURE 3.10
Tea data: representation of confidence ellipses for a few variables.

It is also possible to construct confidence ellipses for the categories of a single categorical variable. To do so, we keep only one variable and suppress the labels of the individuals (see Figure 3.11):

```
> res.mca <- MCA(tea,quanti.sup=22,quali.sup=c(19:21,23:36),graph=FALSE)
> plotellipses(res.mca, keepvar=11, label="none"))
```

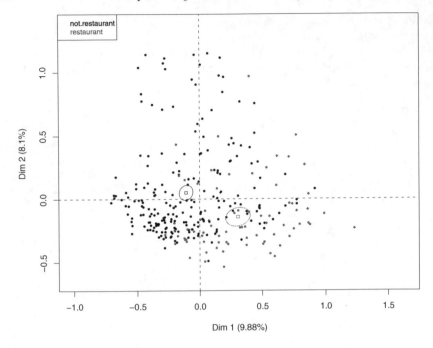

FIGURE 3.11
Tea data: representation of confidence ellipses around the categories of the
variable *Restaurant*.

3.7   Addendum

3.7.1   Analysing a Survey

3.7.1.1   Designing a Questionnaire: Choice of Format

When writing a questionnaire, researchers often want to use multiple choice
questions. By design, these questions can yield a different number of responses
to each question depending on the participant. In the example, the question
originally posed regarding the image of tea was: "Which of the following words
would you associate with the image of tea?" Participants can therefore choose
from the following list: exotic, spiritual, good for you, diuretic, friendliness,
absorb iron, feminine, refined, slimming, stimulant, relaxant, no effect on
health. To use this information, each word is effectively considered as a binary
question ("Do you consider tea to be exotic? *yes/no*"). This multiple choice
question therefore becomes 12 binary questions. In terms of the data table,
we would therefore have one column (one variable) per word.

It is also possible to process data collected using open-ended questions, that is, for which there are no proposed answers. In the example, tea drinkers answered the following question: "Why do you drink tea?" In this example, the words used were listed and those which were frequency cited were selected. From this list, binary questions were created for each word. If the participant cited the word, we attributed the category *yes*, and if not, we attributed *no*. The open question is therefore processed in the same way as the other binary questions. This approach can give rise to a great number of binary variables and therefore lead to the individuals being represented in larger spaces. Among other things, the category *yes* for these variables generally has a very low frequency, and introducing them as active variables is only very rarely satisfactory. In a case such as this, it may prove interesting to group the words together according to meaning (lemmatisation, see Section 2.7). It is preferable, however, not to use too many open questions.

When we wish to consider a quantitative variable as active, it is possible to recode this variable into categories in order to make it categorical. There are a number of different choices in terms of recoding: categories with equal-width bins, categories with equal-count bins, categories according to naturally occurring divisions (these divisions can be visualised on a histogram or obtained automatically using a clustering method, see Section 4.11).

In the case of questions being conditioned by the answers to a previous question j (known as overlapping questions), one way of analysing the data is to consider each of the subpopulations brought about by each category of j. In the example, the question "Do you drink tea?" divided our participants in half and we were only interested in tea drinkers. If we had studied all of our participants, the first components of the MCA would simply have opposed tea drinkers and non-tea drinkers, insomuch as non-tea drinkers systematically answer *no* to the different locations, different times of day, and so forth. It is therefore preferable to limit the number of these types of questions.

It must finally be noted that the number of categories can differ from one variable to the next: those variables which have more categories have greater inertia, but this inertia is shared over a greater number of components. Therefore, the first dimensions will be made up of those variables with very few categories, and those with many.

3.7.1.2 Accounting for Rare Categories

When certain variables admit categories with small sample sizes, there are a number of ways to ensure that these categories do not have too much influence on the analysis.

- Naturally grouping certain categories. This solution is recommended for sequential categories: for example, grouping the categories *70–85 years* and *85 years and over*.

- Ventilation. The concept of ventilation is to randomly attribute those individuals associated with rare categories to other categories. To do so, the proportions of the other categories are calculated and used when attributing those individuals with rare categories.

- Elimination of individuals with rare categories. This solution should be avoided wherever possible. It should only be used if all of the rare categories are due to a very small number of individuals (situation which sometimes occurs when questions remain unanswered).

3.7.2　Description of a Categorical Variable or a Subpopulation

Multidimensional analysis is often supplemented by univariate analyses which are used to characterise a number of specific variables. We shall here focus on describing a specific categorical variable as well as groups of individuals defined by the categories of this variable. To do so, we can use quantitative variables, categorical variables, or the categories of categorical variables.

For example, we shall here describe the variable *type* in detail (cheapest, luxury, supermarket, etc.); one interesting feature of this variable is that it has more than two categories. The results of the **catdes** function applied to the variable *type* are detailed as follows:

```
> catdes(tea,num.var=18)
```

3.7.2.1　Description of a Categorical Variable by a Categorical Variable

To evaluate the relationship between the categorical variable we are interested in (*type*), and another categorical variable, we can conduct a χ^2 test. The smaller the p-value associated with the χ^2 test, the more questionable the independence hypothesis, and the more the categorical variable characterises the variable *type*. The categorical variables can therefore be sorted in ascending order of p-value. In the example (see Table 3.5), the variable *place of purchase* is the most closely related to the variable *type*.

TABLE 3.5
Tea Data: Description of the Variable *Type* by
the Categorical Variables (Object `$test.chi2`)

	p-value	Df
Place of purchase	1.1096e-18	10
Format	8.4420e-11	10
Tearoom	1.6729e-03	5
Friends	4.2716e-02	5
Slimming	4.3292e-02	5
Variety	4.9635e-02	10

3.7.2.2 Description of a Subpopulation (or a Category) by a Quantitative Variable

For each category of the categorical variable *type* and for each quantitative variable (denoted X), the v-test (a test-value) is calculated as follows:

$$\text{v-test} = \frac{\bar{x}_q - \bar{x}}{\sqrt{\frac{s^2}{I_q}\left(\frac{I-I_q}{I-1}\right)}},$$

where $\bar{x}_q$ is the average of variable X for the individuals of category q, $\bar{x}$ is the average of X for all of the individuals, and I_q is the number of individuals carrying the category q. This value is used to test the following null hypothesis: *the values of X for the individuals who chose the category q are selected at random from all of the possible values of X.* We therefore consider the random variable $\bar{X}_q$, average of the individuals for category q. Its expected value and variance are

$$\mathbb{E}(\bar{X}_q) = \bar{x} \quad \text{and} \quad \mathbb{V}(\bar{X}_q) = \frac{s^2}{I_q} \times \frac{I-I_q}{I-1}.$$

The v-test can therefore be considered a "standardised" deviation between the mean of those individuals with the category q and the general average. Among other things, we can attribute a probability to the v-test. If, among the participants, X is normally distributed according to the null hypothesis, the $\bar{X}_q$ distribution is as follows:

$$\bar{X}_q = \mathcal{N}\left(\bar{x}, \frac{s}{\sqrt{I_q}}\sqrt{\frac{I-I_q}{I-1}}\right).$$

If X is not normally distributed, we can still use normal distribution as an approximate distribution for $\bar{X}_q$. We consider the v-test as a statistic of the test for H_0 ("the average of X for category q is equal to the general average," or in other words, "variable X does not characterise category q") and can therefore calculate a *p*-value.

Remark
When categories stem from a clustering, this test can only be applied satisfactorily to supplementary variables (i.e., which were not used to determine the categories), but they are also calculated for the active variables for information.

As the *p*-value provides an indication of the "significance" of a given deviation, it makes sense to organise the quantitative variables in descending order of v-test by limiting oneself to *p*-values less than 5%.

In the example (see below), the only category to be characterised by a quantitative variable is *t_luxury*. This category is characterised by individuals of above-average age as the v-test is positive. The average age of those who buy in this class is 43.4 years whereas the average overall age is 37.1 years.

The standard deviations are provided for both the class (16.9) and the overall population (16.8).

```
> catdes(tea,num.var=18)
$quanti$cheapest
NULL

$quanti$known.brand
NULL

$quanti$luxury
      v.test Mean in category  Overall mean  sd in category  Overall sd  p.value
age   3.02                43.4         37.1            16.9        16.8  0.00256

$quanti$shop.brand
NULL

$quanti$unknown
NULL

$quanti$varies
NULL
```

3.7.2.3 Description of a Subpopulation (or a Category) by the Categories of a Categorical Variable

The description of a categorical variable can be refined by studying the relationships between categories. We thus characterise each of the categories of the variable we are interested in (variable *type*) by using the categories of the categorical variables.

These calculations are illustrated using first the variable *place of purchase* and second the contingency table for the variables *type* and *place of purchase* (see Table 3.6).

TABLE 3.6
Tea Data: Contingency Table for the Variables *Type* and *Place of Purchase*

	Supermarket	Supermarket and specialist	Specialist	Total
Cheapest	6	1	0	7
Luxury	12	20	21	53
Unknown	10	1	1	12
Famous brand	82	11	2	95
Shop brand	20	1	0	21
Varies	62	44	6	112
Total	192	78	30	300

Let us examine the category *luxury* and consider the variable *place of purchase* which has three categories: *supermarket*, *supermarket+specialist* and *specialist shop*. We shall look more closely at *specialist shop* (see Table 3.7). The following question is raised: "Is the category *luxury* characterised by the category *specialist shop*?" The objective is to calculate the proportion of individuals who buy their tea in a *specialist shop* out of those who buy luxury tea I_{qt}/I_q from the overall percentage of individuals who buy their tea in specialist shops I_t/I.

TABLE 3.7
Tea Data: Specialist Shop and Luxury

	Specialist shop	Other	Total
Luxury	$I_{qt} = 21$	32	$I_q = 53$
Other	9	238	247
Total	$I_t = 30$	270	$I = 300$

These two proportions are equal under the null hypothesis of independence:

$$\frac{I_{qt}}{I_q} = \frac{I_t}{I}.$$

I_q individuals are randomly selected (those with the category we are interested in *luxury*) among I (the total population). We shall focus on the random variable X equal to the number I_{qt} of occurrences of individuals which have the characteristic that is being studied (purchased in a specialist shop), while it must be remembered that their sample size within the population is I_t. Under the null hypothesis, the random variable X follows the hypergeometric distribution $\mathcal{H}(I, I_t, I_q)$. The probability of having a more extreme value than the observed value can therefore be calculated. For each category of the variable being studied, each of the categories of the characterising categorical variables can be sorted in ascending order of p-value.

The first row of Table 3.8 indicates that 70% (21/30; see Table 3.6 or the extract) of the individuals who buy their tea in specialist shops also belong to the class *luxury*; 39.6% (21/53; see Table 3.6) of the individuals from the class *luxury* purchase their tea in specialist shops; 10% (30/300; see Table 3.6) of the participants purchase their tea in specialist shops. The p-value of the test (1.58e-11) is provided along with the associated v-test (6.64). The v-test here corresponds to the quantile of the normal distribution which is associated with p-value; the sign indicates an over- or underrepresentation (Lebart et al., 2006).

The categories of all the categorical variables are organised from most to least characteristic when the category is overrepresented in the given class (i.e., the category in question) compared to the other categories (the v-test is therefore positive), and from least characteristic to most when the category is underrepresented in the class (and the v-test is therefore negative). The individuals who buy luxury tea are most significantly characterised by the fact that they do not buy tea in supermarkets (the v-test for supermarkets is negative, and has the highest absolute value).

3.7.3 The Burt Table

A Burt table is a square table of $K \times K$ dimensions, where each row and each column correspond to one of the categories K of the set of variables. In the

TABLE 3.8
Tea Data: Description of the Category *Luxury* of the Variable *Type* by the
Categories of the Categorical Variables (Object $category$luxury)

	Cla/Mod	Mod/Cla	Global	*p*-value	v-test
Place.of.purchase=specialist.shop	70.00	39.6	10.0	3.16e-11	6.64
Format=loose	55.60	37.7	12.0	5.59e-08	5.43
Variety=black	28.40	39.6	24.7	1.15e-02	2.53
Age_Q=60 and +	31.60	22.6	12.7	3.76e-02	2.08
No.effect.health=no.effect.health	27.30	34.0	22.0	3.81e-02	2.07
No.effect.health=not.without.effect	15.00	66.0	78.0	3.81e-02	-2.07
Variety=flavoured	12.40	45.3	64.3	2.86e-03	-2.98
Age_Q=15-24	7.61	13.2	30.7	2.48e-03	-3.03
Format=sachet	8.24	26.4	56.7	1.90e-06	-4.76
Place.of.purchase=supermarket	6.25	22.6	64.0	2.62e-11	-6.67

cell (k, k') we observe the number of individuals who carry both categories k and k'. This table is an extension of the contingency table where there are more than two categorical variables: it juxtaposes all of the information from the contingency table of variables taken as pairs (in rows and columns).

A correspondence analysis of this table is used to represent the categories. As this table is symmetrical, the representation of the cloud of row profiles is identical to that of the cloud of column profiles (only one of the two representations is therefore retained). This representation is very similar to the representation of the categories as provided by MCA and demonstrates the collinearity of the principal components of the same rank. However, the inertias associated with each component differ by a coefficient of λ_s. When λ_s is the inertia of s for the MCA, the inertia of component s for a CA of the Burt table will be λ_s^2. It can be observed that the percentages of inertia associated with the first components of the CA of the Burt table are higher than the percentages of inertia associated with the first components of the MCA alone. In the example, the percentages of inertia associated with the first two components of the MCA are worth 9.88% and 8.10%, respectively, compared with 20.73% and 14.11% for those of the CA.

The Burt table is therefore useful in terms of data storage. Rather than conserving the complete table of individuals × variables, it is in fact sufficient to construct a Burt table containing the same information in terms of associations between categories, which are considered in pairs with a view to conducting the principal component method. When dealing with a very large number of individuals, the individual responses are often ignored in favour of the associations between categories.

3.7.4 Missing Values

It is very common for some data to be missing, for a survey conducted by questionnaire, for example. The easiest way to manage missing values in datasets with categorical variables is to create a new category for each variable which contains one or more missing values. A variable j with K_j categories will therefore have $K_j + 1$ categories if at least one individual possesses missing data for variable j. It is therefore possible to conduct the MCA on the new data table by considering the missing categories in the same way as the other categories.

In practice, this detects the structure of the missing data in the dataset, and enables the structure of the missing data to be viewed. For example, most individuals who did not answer question 8 also left out questions 9 and 10 and this strong structure is highlighted by the first dimension or one of the first dimensions of the MCA.

Viewing the organisation of missing data and understanding why certain individuals did not answer a group of questions can be informative but leads to excessive structuring of the dataset and can conceal the information held within the answers. In some applications, an MCA constructed from datasets with 9% of missing data can have their first two dimensions almost exclusively constructed from associations of missing categories.

If we want to focus on the responses alone, and ignore unanswered questions, it is possible to use an algorithm which functions on the same principle as that of the iterative PCA defined in Section 1.8.4 on page 28. We shall explain the principle of the iterative MCA algorithm. The complete disjunctive table (CDT) is calculated first, and it is this CDT which will be imputed iteratively. An initial phase consists of attributing the missing data from the CDT with a given value (the mean of the category, for example). Then, two phases are alternated: in the first, the MCA is constructed from the previously completed CDT, and in the second the coordinates of the individuals and variables from the MCA are used to reconstitute the data. This reconstitution depends on the number of dimensions in the MCA, which must be defined. When the algorithm converges, an imputed complete disjunctive table is obtained from which we can construct an MCA.

In MCA missing data can be managed using the **imputeMCA** function from the missMDA package, which will provide a completed CDT. The MCA can then be launched using the **MCA** function in FactoMineR applied to the incomplete table using the completed CDT. The lines of code needed to construct an MCA on an incomplete dataset are as follows:

```
> library(missMDA)
> nb <- estim'ncpMCA(donNA)            # choose the number of components
> IndMat <- imputeMCA(donNA, ncp=nb)   # impute the indicator matrix
> res.pca <- PCA(donNA, tab.disj=IndMat$tab.disj)  # perform MCA
```

3.8 Example: The Survey on the Perception of Genetically Modified Organisms

3.8.1 Data Description — Issues

The French, and indeed many people worldwide, are worried about genetically modified organisms (GMOs). On 5 February 2008, these worries were heightened by the announcement by the French agricultural minister, Michel Barnier, that "from 2008" the testing of GMO would again take place in open fields, thus going back on the agreements made at the Environmental Conference. The same year, a trial began against the *faucheurs volontaires* (GMO crop saboteurs) who, the previous year, had destroyed a field of GMO corn grown by Monsanto, a company specialising in agricultural biotechnology.

In light of this situation, two Agrocampus students conducted a study with 135 participants in order to get an overall impression of people's views of GMO. Participants were asked to answer a set of 21 closed questions which were subdivided into two groups.

The first group was made up of 16 questions directly linked to the participants' opinion of GMO:

1. "Do you feel implicated in the debate about GMO (a lot, to a certain extent, a little, not at all)?"

2. "What is your view of GMO cultivation in France (very favourable, favourable, somewhat against, totally opposed)?"

3. "What do you think of the inclusion of GM raw materials in products for human consumption (very favourable, favourable, somewhat against, totally opposed)?"

4. "What do you think of the inclusion of GM raw materials in products to be fed to animals (very favourable, favourable, somewhat against, totally opposed)?"

5. "Have you ever taken part in an anti-GMO protest (Yes/No)?"

6. "Do you think the media communicate enough information about GMO (Yes/No)?"

7. "Do you take it upon yourself to find out more information about GMO (Yes/No)?"

8. "Do you think that GMO might enable us to reduce the use of fungicides (Yes/No)?"

9. "Do you think that GMO might enable us to reduce the problems of hunger in the world (Yes/No)?"

10. "Do you think that the use of GMO might help to improve farmers' lives (Yes/No)?"

11. "Do you think that GMO might lead to future scientific advances (Yes/No)?"

12. "Do you think that GMO represent a danger to our health (Yes/No)?"

13. "Do you think that GMO represent a possible danger to the environment (Yes/No)?"

14. "Do you think that GMO represent a financial risk for farmers (Yes/No)?"

15. "Do you think that GMO are a useless scientific practice (Yes/No)?"

16. "Do you think our grandparents' generation had a healthier diet than us (Yes/No)?"

The second group was made up of five descriptive variables:

1. Sex (male, female)

2. Professional status (farmer, student, manual labourer, senior management, civil servant, accredited professional, technician, retailer, other profession, unemployed, retired)

3. Age (−25 years, 25–40 years, 40–60 years, +60 years)

4. "Is your profession or education in any way linked to agriculture, the food industry or the pharmaceutical industry (Yes/No)?"

5. "Which political movement do you most adhere to (extreme left, green, left, liberal, right, extreme right)?"

Using this questionnaire, we wanted, on the one hand, to characterise our participants in terms of their relationship with GMO, and on the other hand, to see if this characterisation has any relation to the descriptive variables. The question "Is your profession or education in any way linked to agriculture, the food industry or the pharmaceutical industry?" is important when interpreting the results as it is probable that those people who answer yes to this question might have greater scientific knowledge of GMO than the other participants.

The first thing to do is to construct frequency tables for the questions to observe how the responses to each question are distributed. To do so, the following line of code is used to yield the sample sizes for each of the categories of the first 16 variables:

```
> gmo <- read.table("http://factominer.free.fr/bookV2/gmo.csv",
    header=TRUE,sep=";",dec=".",row.names=1)
> summary(gmo[,1:16])
 Implicated        Position.Culture       Position.Al.H
 A little      :31  Favourable      :45    Favourable      :37
 A lot         :36  Somewhat Against:54    Somewhat Against:47
 Certain extent:53  Totally opposed :33    Totally opposed :50
 Not at all    :15  Very Favourable : 3    Very Favourable : 1
```

```
Position.Al.A          Protest   Media.Passive Info.Active
Favourable      :44    No :122   No :78        No :82
Somewhat Against:39    Yes: 13   Yes:57        Yes:53
Totally opposed :44
Very Favourable : 8


Phytosanitary.products Hunger    Animal.feed Future.Progress Danger   Threat
No :56                 No :67    No :93      No :54          No :39   No :48
Yes:79                 Yes:68    Yes:42      Yes:81          Yes:96   Yes:87


Finan.risk Useless.practice Grandparents
No :67     No :123          No :49
Yes:68     Yes: 12          Yes:86
```

The summary of the active dataset led us to group together certain categories due to their limited number (see Section 3.7.1.2). In answer to the question "What do you think of the inclusion of GM raw materials in products for human consumption?" for example, only one person responded *very favourable*. As a result, this category thus has an extremely small sample size and therefore it is preferable to group it together with another category. In this particular case, it is relatively easy to group the category with another as the variable concerned is made up of sequential categories: it would not be considered unreasonable to replace *very favourable* with *favourable*. To do so, the following line of code is used to group the categories *very favourable* and *favourable* together to form one category *favourable*:

```
> levels(gmo$Position.Al.H)[4] <- levels(gmo$Position.Al.H)[1]
```

Similarly, for the question "What is your view of GMO cultivation in France?" we group the categories *very favourable* and *favourable* together to form one category *favourable*.

```
> levels(gmo$Position.Culture) <- c("Favourable","Somewhat Against",
"Totally opposed","Favourable")
```

Once the categories have been recoded, the summary of the dataset is as follows:

```
> summary(gmo[,1:16])
       Implicated         Position.Culture        Position.Al.H
A little       :31    Favourable      :48    Favourable      :37
A lot          :36    Somewhat Against:54    Somewhat Against:47
Certain extent:53     Totally opposed :33    Totally opposed :50
Not at all     :15                           Very Favourable : 1


           Position.Al.A Protest   Media.Passive Info.Active
Favourable      :44    No :122   No :78        No :82
Somewhat Against:39    Yes: 13   Yes:57        Yes:53
Totally opposed :44
Very Favourable : 8


Phytosanitary.products Hunger    Animal.feed Future.Progress Danger   Threat
No :56                 No :67    No :93      No :54          No :39   No :48
```

```
Yes:79                    Yes:68   Yes:42        Yes:81          Yes:96   Yes:87

Finan.risk Useless.practice Grandparents
No :67     No :123          No :49
Yes:68     Yes: 12          Yes:86
```

Generally speaking, when the categories of a given question are unremarkable (if there is no relationship between them), the category that is rarely used can be replaced by another chosen at random from the remaining, frequently used categories.

The following line of code yields the frequency tables for the descriptive variables:

```
> summary(gmo[,17:21],maxsum=Inf)
 Sex          Age                   Profession Relation     Political.Party
 F:71   [26; 40]:24   Accredited Prof   : 3    No :79    Extreme left: 9
 H:64   [41; 60]:24   Civil Servant     : 9    Yes:56    Greens      : 7
        < 25    :73   Manual Labourer   : 1              Left        :47
        > 60    :14   Other             : 9              Liberal     :32
                      Retailer          : 3              Right       :40
                      Retired           :14
                      Senior Management :17
                      Student           :69
                      Technician        : 6
                      Unemployed        : 4
```

The following section will show that it is not necessary to group together the categories for these data.

3.8.2   Analysis Parameters and Implementation with FactoMineR

In light of the objectives outlined in Section 3.1, it is natural to describe the individuals according to their responses to the first 16 questions; those relating to their opinion of GMO. The first 16 questions will therefore be considered active variables, and the next 5 questions as illustrative variables. By design, the illustrative variables do not contribute to the construction of the principal components and likewise for their associated categories. Therefore, it is unnecessary to group together the rarer categories.

The following line of code is used to conduct such an analysis:

```
> res <- MCA(gmo,ncp=5,quali.sup=17:21,graph=FALSE)
> summary(res)
> res
**Results of the Multiple Correspondence Analysis (MCA)**
The analysis was performed on 135 individuals, described by 21 variables
*The results are available in the following objects:

   name                description
1  "$eig"              "eigenvalues"
2  "$var"              "results for the variables"
```

```
3  "$var$coord"        "category coordinates"
4  "$var$cos2"         "cos2 for the categories"
5  "$var$contrib"      "contributions of the categories"
6  "$var$v.test"       "v-test for the categories"
7  "$ind"              "results for the individuals"
8  "$ind$coord"        "individuals' coordinates"
9  "$ind$cos2"         "cos2 for the individuals"
10 "$ind$contrib"      "contributions of the individuals"
11 "$quali.sup"        "results for the supplementary categorical variables"
12 "$quali.sup$coord"  "coord. for the supplementary categories"
13 "$quali.sup$cos2"   "cos2 for the supplementary categories"
14 "$quali.sup$v.test" "v-test for the supplementary categories"
15 "$call"             "intermediate results"
16 "$call$marge.col"   "weights of columns"
17 "$call$marge.li"    "weights of rows"
```

It is also possible to group the categories automatically using ventilation, as described in Section 3.7.1.2. This distribution is random or accounts for the order of the categories within a variable if the variable is ordered (**ordered** in R). To automatically group the categories, the following line of code can be used:

```
> res <- MCA(gmo,ncp=5,quali.sup=17:21,graph=FALSE,level.ventil=0.05)
```

where **level.ventil** designates the threshold below which a category is ventilated. In the example, if a category is selected by less than 5% of individuals, they are distributed among the existing categories.

3.8.3 Analysing the First Plane

To visualise the cloud of individuals, the following line of code is used:

```
> plot.MCA(res,invisible=c("var","quali.sup"),label=FALSE)
```

The cloud of individuals on the first plane (see left graph, Figure 3.12) is shaped like a parabola: this is known as the Guttman effect (or horseshoe effect). This effect illustrates the redundancy of the active variables, or in other words, a cloud of individuals that is highly structured according to the first principal component. In the example, this is represented, on the one hand, by two extreme positions relating to GMO distributed on both sides of the first principal component, and on the other hand, by a more moderate position, situated along the length of the second principal component. No further conclusions can be drawn by simply looking at the cloud of individuals, which must be interpreted in conjunction with the cloud of categories.

To visualise the cloud of individuals, the following line of code is used:

```
> plot.MCA(res,invisible=c("ind","quali.sup"),label=FALSE)
```

In the same way as the cloud of individuals, the shape of the cloud of categories on the first plane (see right graph, Figure 3.12 or Figure 3.13) resembles a parabola, and thus still represents a Guttman effect.

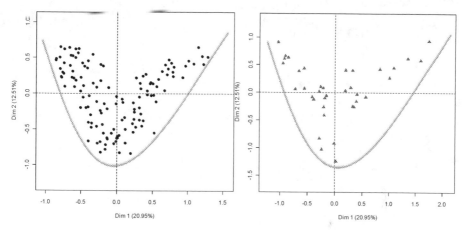

FIGURE 3.12
GMO data: representation of individuals (left) and active categories (right) on the first plane.

To interpret the principal component, it is essential to represent the categories in association with their labels, which is done using the following line of code:

```
> plot.MCA(res,invisible=c("ind","quali.sup"))
```

On the positive side of the first principal component (Figure 3.13), we can observe those people who feel implicated by the debate surrounding GMO and who are somewhat against their use (through the categories they chose). On the negative side of the same principal component, we can see those people who do not feel implicated by the debate surrounding GMO and who are in favour of their use.

Along the second principal component, we can also observe those people with less distinct opinions who feel somewhat implicated by the debate surrounding GMO and who are somewhat against their use.

3.8.4 Projection of Supplementary Variables

It may now be interesting to examine whether or not the structure observed for the individuals concerning their position in terms of GMO can be linked with who they are. In other words, can the relationship to GMO be explained by the descriptive data? To find out, the cloud of illustrative categories is visualised on the first plane using the following line of code:

```
> plot.MCA(res,col.quali.sup="brown",invisible=c("quanti.sup","ind","var"))
```

This representation of supplementary variables (see Figure 3.14) is particularly remarkable as it provides two types of information. First, it reveals a

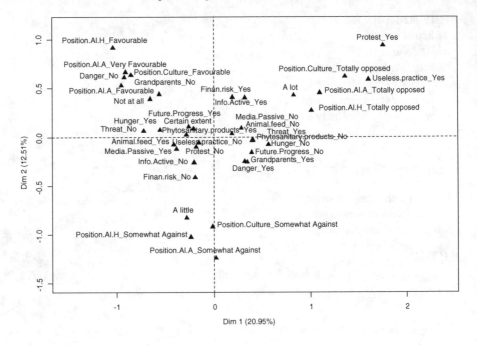

FIGURE 3.13
GMO data: representation of active categories and their labels on the first plane.

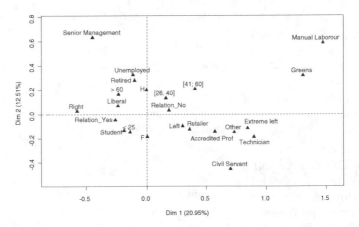

FIGURE 3.14
GMO data: representation of illustrative categories and their labels on the first plane.

strong structure for both of the variables *profession* and *identification with a political movement*, and second, it fails to identify any particular structure with the variables of age, sex, or profession in relation to agriculture, the food industry, and the pharmaceutical industry.

The categories *senior management*, *unemployed*, and *retired* are in opposition to the categories *technician* and *manual labourer* to *civil servant* between the two groups. Similarly, the category *right* is opposed to the categories *green* and *extreme left*, to, in the middle, *left*.

3.8.5 Conclusion

The relationships between these three clouds of points suggest three very different positions in terms of GMO. These positions must be compared with both the professions in the survey and the political movement with which the participant identifies himself. These two variables seem to be strongly linked. However, these three positions do not seem to be explained by sex, age, or whether or not the participant's profession has a link with the industry, which would provide further scientific knowledge of GMO.

3.9 Example: The Sorting Task Dataset

3.9.1 Data Description — Issues

Classification is a cognitive process by which different objects are grouped together by a set of subjects according to their similarities. It is sometimes referred to as a holistic approach because the objects to be categorised are considered as a whole. Classification is used to collect data, particularly in sensory analysis, where the aim is to understand a set of products according to their sensory properties. In this specific context, this task means asking consumers/subjects/a jury to group products together according to their sensory similarities. This section presents a somewhat unusual application of MCA to data so unique that each variable may be considered as a section of a set of objects. This process will be outlined below.

The data which we will be using come from a body of sensory data collected at Agrocampus. Ninety-eight consumers conducted a categorisation task using 12 luxury perfumes: Angel, Aromatics Elixir, Chanel 5, Cinéma, Coco Mademoiselle, J'adore (eau de parfum), J'adore (eau de toilette), L'instant, Lolita Lempicka, Pleasures, Pure Poison, and Shalimar (the labels for each perfume were, of course, hidden). The participants were asked to divide the perfumes into groups according to their sensory similarities, and then to attribute a description to each of the groups.

First, the data are organised into a table with 12 rows and 98 columns,

TABLE 3.9
Perfume Data: Categorisation by Judges 18, 31, 40, and 93

	Judge 18	Judge 31	Judge 40	Judge 93
Angel	1	1	6	1
Aromatics Elixir	2	2	5	2
Chanel 5	1	3	5	2
Cinéma	1	4	3	3
Coco Mademoiselle	1	5	2	3
J'adore (eau de parfum)	3	5	1	1
J'adore (eau de toilette)	1	5	1	3
L'instant	2	5	2	1
Lolita Lempicka	3	4	3	3
Pleasures	1	5	1	1
Pure Poison	3	2	2	3
Shalimar	2	5	4	4

in which each row i corresponds to a perfume, each column j corresponds to a consumer, and each cell (i, j) corresponds to the group into which the product i was grouped by the consumer j (see Table 3.9). Each consumer j can therefore be assimilated with a categorical variable j with K_j categories, where K_j designates the number of groups used by the consumer j when categorising the data. For example, in Table 3.9, we can see that participant 31 ($j = 31$) separated the perfumes into 5 categories ($K_{31} = 5$) and that s/he categorised J'adore (eau de parfum) and J'adore (eau de toilette) in the same group.

Second, in the same way, the index of the group into which product i was categorised by consumer j can be replaced by the words which categorise that same group. Similarly, each consumer j can be assimilated with a categorical variable of K_j categories (see Table 3.10). We therefore obtain an identical table, but one which is encoded more explicitly. This second table is analysed further below.

TABLE 3.10
Perfume Data: Example of Sorting Task Data with Comments

	Judge 18	Judge 31	Judge 40	Judge 93
Angel	lively	strong	Gr6	strong flowery
Aromatics Elixir	grandmother	spicy	Gr5	unnatural
Chanel 5	lively	soapy	Gr5	unnatural
Cinéma	lively	Gr4	Gr3	weak flowery
Coco Mademoiselle	lively	soft	Gr2	weak flowery
J'adore (eau de parfum)	soft, nice, baby	soft	shower gel	strong flowery
J'adore (eau de toilette)	lively	soft	shower gel	weak flowery
L'instant	grandmother	soft	Gr2	strong flowery
Lolita Lempicka	soft, nice, baby	Gr4	Gr3	weak flowery
Pleasures	lively	soft	shower gel	strong flowery
Pure Poison	soft, nice, baby	spicy	Gr3	weak flowery
Shalimar	grandmother	soft	lemongrass	strong

One of the main aims of this study was to provide an overall image of the 12 luxury perfumes based on the categorisations provided by the 98 consumers. Once this image is obtained, the image's sensory data must be linked to the terms used to categorise the groups to understand the reasons why two perfumes are not alike. Finally, to analyse the data in further detail, we will see how in this specific sensory context, it might be possible to use the barycentric properties of the MCA to make the most of our data.

3.9.2 Analysis Parameters

In this study, the 12 perfumes are considered as (active) statistical individuals, and the 98 participants as (active) categorical variables. A data table with individuals in rows and categorical variables in columns is used and is therefore suitable for a multiple correspondence analysis. It must be noted that, in the analysis, the data are accounted for by the indicator matrix which here has $I = 12$ rows and $K = \sum K_j$ columns: participant j is represented by the set of his/her corresponding K_j variables, with each variable corresponding to a group, and having the value 1 when the perfume belongs to group k and 0 if it does not (see Section 3.4). The distance between two perfumes is such that

1. Two perfumes i and l are merged if they were grouped together by all of the participants.

2. Two perfumes i and l are considered close if they were grouped together by many of the participants.

3. Two products are considered far apart if many of the participants assigned them to two different groups.

To conduct the MCA, the following line of code is used, which stores the results of the MCA in `res.perfume`:

```
> perfume <- read.table("http://factominer.free.fr/bookV2/perfume.csv",
    header=TRUE,sep=";",row.names=1)
> res.perfume <- MCA(perfume)
> summary(res.perfume)
```

By default, this function takes all of the active variables and therefore the only input setting required is the name of the dataset.

3.9.3 Representation of Individuals on the First Plane

To visualise the cloud of individuals, the following line of code is used:

```
> plot.MCA(res.perfume,invisible="var",col.ind="black")
```

The first principal component identifies the perfumes Shalimar, Aromatics Elixir, and Chanel 5 as being opposed to the other perfumes (see Figure 3.15). The second principal component opposes Angel, Lolita Lempicka, and, to a lesser extent, Cinéma, with the other perfumes. The distancing of a small number of perfumes is related to the number of times that these perfumes were

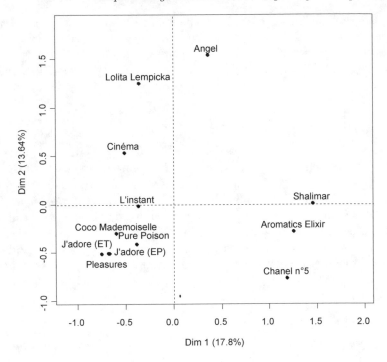

FIGURE 3.15
Perfume data: representation of perfumes on the first plane.

put into groups of their own: this is indeed the case for Shalimar, Chanel 5, and Angel, which were singled out by 24, 17, and 13 participants, respectively. The proximity of certain perfumes is related to the frequency with which they were assigned to the same group: this was the case for Aromatics Elixir, which was associated with Shalimar (Chanel 5) 42 times (and 51 times, respectively); and for Lolita Lempicka, which was associated with Angel 36 times. The two J'adore perfumes were grouped together 56 times and are therefore also close to one another.

3.9.4   Representation of Categories

The representation of the perfumes is supplemented by superimposing the representation of the categories – referred to hereafter as words. By default, a perfume is located at the barycentre of the words with which it was associated. To visualise the cloud of categories and to interpret the oppositions between the perfumes, the following line of code is used:

```
> plot.MCA(res.perfume,invisible="ind",col.var="black")
```

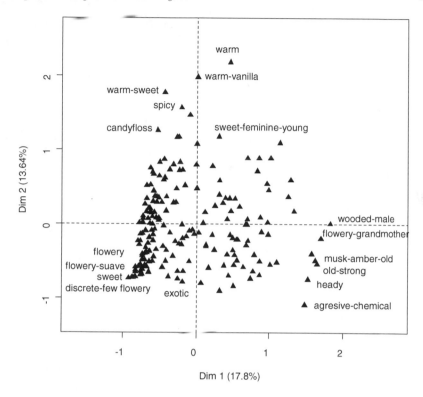

FIGURE 3.16
Perfume data: representation of words on the first plane.

Due to the great number of words, the cloud of categories, as provided by the function `plot.MCA`, cannot be processed directly. Figure 3.16 is a simplified representation of this cloud.

The first component opposes the perfumes associated with the words *old*, *strong*, and the perfumes described as *flowery*, *soft*. The second component opposes the perfumes associated with the words *warm*, *sweet*, and *spicy* with the other perfumes (see Figure 3.16).

3.9.5 Representation of the Variables

The variables can be represented by calculating the correlations between the individuals' coordinates on one component and each of the categorical variables (see Section 3.4.3). In the example, each participant is represented by a point and two participants are close to one another if they categorised the perfumes in the same manner.

Figure 3.17 illustrates different types of categorisations. On the first component, participants 93 and 40 have high coordinates compared with participants 18 and 31. As the participants' coordinates on a component are equal to the correlation between their partitioning variables and the component, participants 40 and 93 clearly singled out the perfumes Shalimar, Aromatics Elixir, and Chanel 5, unlike participants 18 and 31 (see Table 3.9). According to the first component, participants 31 and 40, who have high coordinates, are opposed to participants 18 and 93. On closer inspection of the data, participants 31 and 40 did indeed single out Angel, and to a lesser extent Lolita Lempika and Cinéma, whereas participants 18 and 93 did not (see Table 3.9).

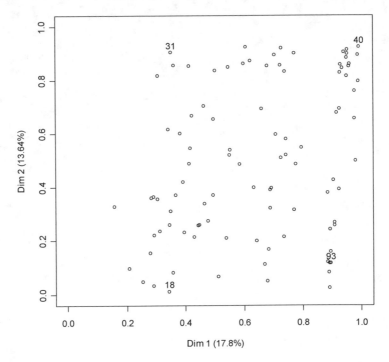

FIGURE 3.17
Perfume data: representation of participants on the first plane.

4

Clustering

4.1 Data — Issues

Multidimensional data analysis (MDA) methods mainly provide synthetical representations of objects (most of the time these objects are individuals, variables, or categories of categorical variables) corresponding to the rows and columns of a data table. In MDA, the most typical method for representing a set of objects is a cloud of points (each point is an object), evolving in a Euclidean space (that can be reduced to a plane representation). The term "Euclidean" here refers to the fact that the distances between points (respectively the angles for the quantitative variables) are interpreted in terms of similarities for the individuals or categories (respectively in terms of correlation for the quantitative variables). Principal component methods, such as principal component analysis (PCA), correspondence analysis (CA), and multiple correspondence analysis (MCA), all yield Euclidean representations.

Another means of representing a set of objects and illustrating the links between them (similarities or correlations) is with a hierarchical tree (see Figure 4.1). This concept is also referred to more simply as a hierarchy, and more precisely, an indexed hierarchy, which reminds us that the levels at which the objects are grouped together can be interpreted (we also use the term "dendrogram"). The way that these trees are used is relatively simple: two objects are similar to one another if, to go from one to the other, we do not need to go too far back up the tree. Thus, in Figure 4.1

- Objects A and B are more similar than objects D and E.

- Object C is more similar to the set of objects D, E than to the set A, B.

It must be noted that the structure of a tree is not modified by conducting symmetries, as shown by the two representations in Figure 4.1. The lateral proximities (between B and C on the left of Figure 4.1, for example) between objects should not be interpreted. In this respect, there is some freedom in the representation of a tree that we can use when a variable plays an important role and individuals are sorted according to that variable. Wherever possible, the branches of each node are switched so that this order can be respected.

The best example of a hierarchical tree is undoubtedly that of living beings, with the first node separating the animal kingdom from plant life. This

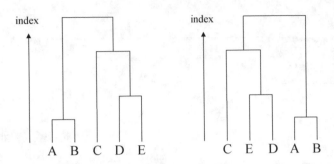

FIGURE 4.1
Example of a hierarchical tree (summarising the similarities between five objects: A, B, C, D, E).

hierarchy is used by all naturalists. Other similar hierarchies are those used to describe the organisation of companies and administrations, which enable everyone to know where they stand within the organisation. Another example is that of the family tree describing the relations between ancestors. Finally, Figure 4.2 is a good example of the synthetical vision provided by a hierarchical tree.

In these examples, the trees were constructed by experts according to the rules usually used to represent that kind of organisation. For example, to represent living creatures, we refer to evolution. Ideally, the different nodes of the tree each represent a stage of evolution, where the most important steps correspond to the nodes at the top of the tree (which for the animal kingdom might first separate single-cell organisms from multicellular organisms). The idea of evolution can therefore be found in the lateral order of individuals: the branches of each node are organised by placing the least evolved animals on the left.

As a result, one criterion (at least) is associated with each node and all of the individuals on a branch coming from this node present the same value for this (or these) criterion (or criteria). A set of individuals such as this is said to be monothetic (i.e., single-celled organisms, invertebrates, mammals, etc.). In this example in particular, but also to a lesser extent in the previous examples, the hierarchical tree is the result of a great deal of research which has led to a value attributed to the most important criteria and has thus enabled researchers to define the nodes at the highest levels.

In this chapter, we will adopt the approach used in principal component methods, and more explicitly, the analysis of a data table without prior judgements (although assumptions are made when constructing the table in choosing the individuals and variables). The aim is to construct a hierarchical tree (rather than a principal component map) in order to visualise the links between the objects, which is a means of studying the variability within the table. This raises the same issues as principal component methods: the only difference between the two approaches is the way they are represented.

Without prior judgements, we shall attempt to construct a hierarchical tree in which each branch groups together the individuals of a polythetic group (groups like this are defined by a set of properties such as, first, each element within the group possesses a great number of these properties, and second, each property is possessed by a great number of individuals within the group).

The algorithms used to construct trees such as these are known as hierarchical clustering. There are a great many of these kinds of algorithms: the most common work in an agglomerative manner (first grouping together the most similar objects and then the groups of objects) and are grouped together under the umbrella term Agglomerative Hierarchical Clustering (AHC). Other algorithms are divisive, i.e., in a touchdown approach. This chapter mainly describes and illustrates one of the most widely used of these AHCs: Ward's algorithm.

A third synthetical representation of the links between objects is the partition obtained by dividing the objects into groups so that each object belongs to one group and one only. In formal terms, a partition can be assimilated to a categorical variable (for which the value for each object is the label — or number — of the group to which it belongs). Thus, in an opinion poll, we will distinguish between men and women, for example, as well as between those who use a given product and those who do not. However, these monothetic groups are only of interest if the partition that they belong to is linked to a great number of variables. In other words, in an opinion poll, the divide between men and women is only really of interest if the responses to the opinion poll are different for men and women, and indeed increases as the number of different responses increases.

Again, here, we are approaching the data in an exploratory manner: from a rectangular data table, we aim to divide up these objects so that, first, the individuals within each class are similar to one another, and second, the individuals differ from one class to the next. Many algorithms are available, all grouped together under the term partitioning; in this chapter we shall limit ourselves to the mostly widely used: K-means algorithm.

In an attempt to keep things general, until now we have referred simply to objects, be they statistical individuals or quantitative variables or categories of categorical variables. Indeed, one of the strengths of clustering methods is that their general principles can be applied to objects with various features. However, this generality impairs the tangible nature of our presentation. Similarly, in the following sections, we shall restrict our presentation to objects corresponding to statistical individuals, described by a set of quantitative or categorical variables: in practice, this case is frequently encountered.

Classifying and Classing

Classifying a set of objects means establishing or constructing clusters or a hierarchy. Classing an object means organising this object into one of the groups of a predefined (and unquestionable) partition. This operation is known as

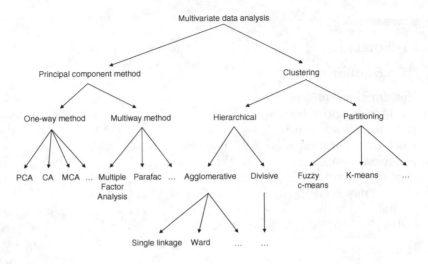

FIGURE 4.2
Hierarchical tree illustrating the links between the principal methods of data
analysis.

clustering. In statistics, the term "discrimination" corresponds to the problem of finding rules for classification (of individuals into one of the groups of
a predefined partition from a set of available variables). Bearing in mind that
a partition can be considered a categorical variable, the problem with discrimination lies in predicting a categorical variable, from quantitative and/or
categorical variables, in the same way that regression methods aim to predict
quantitative variables. The most widely known example of discrimination is
that of medical diagnosis: when someone is ill, we have access to a number
of values for different variables; how do we work out what illness he or she is
suffering from? Each illness is a category of the categorical variable: the set
of categories (one category = one illness) is never questioned.

Supervised and Unsupervised Classification
Researchers have introduced the term "unsupervised" classification to designate that which was, for a long time (including in this work) simply referred
to as clustering. This term is supposed to refer to the exploratory nature of
these methods. This is in contrast to supervised classification, which designates that which was referred (including in this work) to as discrimination.
The term "supervised" is supposed to refer to the way in which the approach
focuses on a variable (the categorical variable to be predicted).

4.2 Formalising the Notion of Similarity

4.2.1 Similarity between Individuals

In clustering, be it hierarchical or otherwise, it is essential to specify what is meant by similarities between two individuals. This requirement also exists in principal component methods but is less evident as the specification is included in the method. In contrast, in clustering, one has the choice, which is an advantage when faced with unusual data.

4.2.1.1 Distances and Euclidean Distances

In the case of standardised PCA, when I individuals are described by K quantitative variables (an individual can be assimilated to an element of the vectorial space $\mathbb{R}^K$), the similarity between two individuals i and l is obtained using the usual (Euclidean) distance in $\mathbb{R}^K$:

$$d^2(i,l) = \sum_{k=1}^{K} (x_{ik} - x_{lk})^2,$$

$$d^2(i,l) = \sum_{k=1}^{K} \left(\frac{x_{ik} - \bar{x}_k}{s_k} - \frac{x_{lk} - \bar{x}_k}{s_k} \right)^2,$$

$$d^2(i,l) = \sum_{k=1}^{K} \frac{1}{s_k}(x_{ik} - x_{lk})^2,$$

where x_{ik} denotes the value of individual i for variable k, $\bar{x}_k$ (and s_k, respectively) the average (and the standard deviation, respectively) of variable k. In the first more general equation, the data x_{ik} were first centred and reduced. In the other two, the centring and reduction are explicitly detailed, and will be explained later in this chapter.

When defining the distance d between individuals within a space (here $\mathbb{R}^K$), we say that this space is attributed the distance d (also known as the "metric space"). The d function from $I \times I$ to $\mathbb{R}^+$ possesses all of the desired mathematical properties of a distance

$$\begin{cases} d(i,l) &= 0 \Longleftrightarrow i = l, \\ d(i,l) &= d(l,i), \\ d(i,l) &\leq d(i,j) + d(j,i) \ \text{(triangle inequality)}. \end{cases}$$

Among other things, we are dealing with Euclidean distance, that is to say, it can be used to define the notion of angle and thus of orthogonal projection (an axiomatic definition of the concept of Euclidean space goes beyond the framework of this work). As this notion is central to principal component methods, all principal component methods use Euclidean distance (this is also the case for the χ^2 distance in CA, for example). However, if the notion of projection is not required, we are not restricted to using Euclidean distance. This is one of the most important features of these clustering methods in terms of measuring the similarities between individuals.

4.2.1.2 Example of Non-Euclidean Distance

In usual Euclidean distance, the deviations for each variable are squared. The influence of these great deviations is therefore intensified, which is why it can be preferable to instead use the deviations' absolute values. This yields the following distance, between individuals i and l:

$$d(i,l) = \sum_{k=1}^{K} |x_{ik} - x_{lk}|.$$

This distance is known as the Manhattan distance, or even city block distance, in reference to the American towns where all the streets are either parallel or orthogonal: in order to move from one point to another, the distance travelled is equal to the above distance. This distance can be easily interpreted directly from the data.

Figure 4.3 illustrates the difference between the two distances for a rather basic case. In this example, individuals a and b only differ by one variable, but the difference is extreme; individuals a and c (or b and c) differ on a number of variables, although only slightly. For usual Euclidean distance, the distance between a and b is the greatest: $d(a,b) = 2 > \sqrt{3} = d(a,c) = d(b,c)$; the opposite is true for the city block distance.

	V1	V2	V3
a	1	1	3
b	1	1	1
c	2	2	2

A

	a	b	c
a	0		
b	2	0	
c	$\sqrt{3}$	$\sqrt{3}$	0

B

	a	b	c
a	0		
b	2	0	
c	3	3	0

C

FIGURE 4.3
Normal Euclidean distance (B) and city block distance (C) illustrated for three individuals a, b, c described by three variables $V1$, $V2$, $V3$ (A).

City block distance is not a Euclidean distance. So, which one should we choose? Unless required by the data (a situation we have not as yet encountered), we recommend using a Euclidean distance as it means both clustering and principal component methods can be conducted subsequently.

4.2.1.3 Other Euclidean Distances

There are an infinite number of Euclidean distances. The most well known, and the easiest to interpret, consists of taking the usual distance and attributing a weight to each dimension (i.e., variable). For instance, in standardised PCA, we can assume that the data is merely centred, and that the distance used attributes each variable with a weight opposite to that of its standard deviation (see the third equation defining $d^2(i, l)$, Section 4.2.1.1). Interestingly, these differences illustrate the fact that, when encountering a Euclidean distance, we can also refer back to the usual distance by transforming the data.

4.2.1.4 Similarities and Dissimilarities

Among the first tables to have been used for automatic clustering were those known as presence–absence tables, used in phytosociology. In a given zone, we define a set of stations representative of the diversity of the environments found in the zone; for each station, the different plants present are listed. These data are brought together in a table confronting the species I and the stations J, with the general term x_{ij} having a value of 1 if the species i is present in the station j, and 0 if it is not.

One of the general objectives of this type of study is to illustrate the associations of plants, that is, sets of species present within the same environments. From this objective stems the idea of species clustering, with two species considered similar when observed in the same station (the stations themselves can also be classified; two stations are similar if they have a lot of species in common). This notion of similarity still remains to be defined.

Phytosociologists quickly became aware that, when evaluating the associations between species, their combined presence is more valuable (more environmentally significant) than their combined absence. For this reason, the decision was made to come up with an ad hoc similarity measurement which would take this aspect into account. Many different measurements were suggested. When these measurements do not verify triangle inequality, they are known as dissimilarities (or similarity indices when the value increases with the number of similar individuals). The first of these solutions was suggested by Jaccard (1901)[1]. By denoting, for a pair of species i and l, n_{++} the number of stations where the two species i and l are present and n_{+-} (respectively n_{-+}) the number of stations where the species i is present and l is not (respectively l is present and i is not), Jaccard's index (of similarity) is expressed as

$$\frac{n_{++}}{n_{++} + n_{-+} + n_{+-}}.$$

This index fails to account for the stations in which neither species is present. This type of approach is applied more generally to presence–absence tables

---

[1] Jaccard P. (1901). *Bulletin de la Société Vaudoise des Sciences Naturelles*, 37, 241–272.

confronting individuals (to be classified) and characters, such that the presence of a character is more important for the user than its absence. The characters may also be seen as categorical variables with two categories, and the MCA framework, and particularly its distance, is appropriate.

There are other cases where the nature of the objects being studied is such that the assigned measurement of similarity is not a distance but rather a dissimilarity. One good example is the relationship between genomes. Without going into too much detail, it means measuring the similarity between sequences of letters of the alphabet {a, c, g, t}. One might consider counting the frequency of each series of n letters in each sequence (which may include multiple values for n), and using the χ^2 distance. However, summarising a sequence for a set of frequencies such as this would not be satisfactory. One might think that the relationships between two genomes A and B would be brought closer together by using the lengths of longer sequences of letters common to both A and B. From these lengths, we can construct an acceptable indicator for geneticists, but which does not possess the properties of distance (even without knowing exactly how these lengths are used by the indicator, which is somewhat technical, we may be led to believe that the triangle inequality would not be verified). Clustering methods are extremely valuable in such cases in order to respect the similarity measure, which is suited to the objects to be classified.

4.2.2   Similarity between Groups of Individuals

To construct a hierarchical tree, the distance or dissimilarity between groups of individuals must be defined. There are a number of options in such a case: we shall only cite those which are the most widely used. Let us consider two groups of individuals A and B. The single linkage between A and B is the distance between the two closest elements in the two clusters A and B. The complete linkage between A and B is the greatest distance between an element in A and an element in B. Figure 4.4 illustrates these definitions.

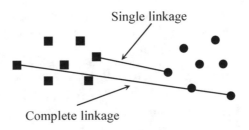

FIGURE 4.4
Single linkage and complete linkage between two groups of individuals (represented by different symbols).

The major advantage of the above definitions is that they can be applied to all distances or dissimilarities. For Euclidean distances, other possibilities are available. Let us consider G_A and G_B the centres of gravity for the sets of individuals A and B. The first option is to measure the dissimilarity between A and B by the distance between their centres of gravity. Another, more satisfactory approach is that of inertia: this means accounting for the groups' weights. In this chapter, individuals are all considered to have the same weight, which is the most common case, and a group's weight is proportional to its size; it must be noted, however, that the inertia approach can simply take into account weights which differ from one individual to another.

Let us apply Huygens' theorem to both A and B ($A \cup B$ with centre of gravity G). Total inertia (of $A \cup B$ with respect to G) = between-clusters inertia (of $\{G_A, G_B\}$ with respect to G) + within-cluster inertia (inertia of A with respect to G_A plus the inertia of B with respect to G_B). This partitioning suggests using between-clusters inertia as a measurement of dissimilarity between A and B. We shall discuss some of the properties of this approach in the section on Ward's method, which is based on this methodology.

4.3 Constructing an Indexed Hierarchy

4.3.1 Classic Agglomerative Algorithm

The starting point in this approach is a dissimilarity matrix D (these dissimilarities might be Euclidean distances) with the general term $d(i, l)$ indicating the dissimilarity between individuals i and l. This matrix is symmetrical, with zeros on the diagonal.

We agglomerate the most similar, or the "closest" individuals i and l (in case of ex-aequo, one couple of individuals is chosen at random), which creates a new element, (i, l). This group of individuals will never be called into question. Value $d(i, l)$ is the agglomerative criterion between i and l. This value is used to determine the height at which the branches of the tree corresponding to i and l are connected.

Matrix D is updated by deleting the rows and columns corresponding to individuals i and l and creating a new row and a new column for the group (i, l) in which the dissimilarities between this group and each of the remaining individuals are noted. We thus obtain matrix $D(1)$ in which we look for the closest pair of elements. These elements are then agglomerated, and so on.

As an example, we shall apply this algorithm to a small dataset with six individuals, as represented in Figure 4.5. To facilitate the calculations, we shall use the city block distance and the complete linkage agglomerative rule. The different stages of the construction of the tree are shown in Figure 4.5.

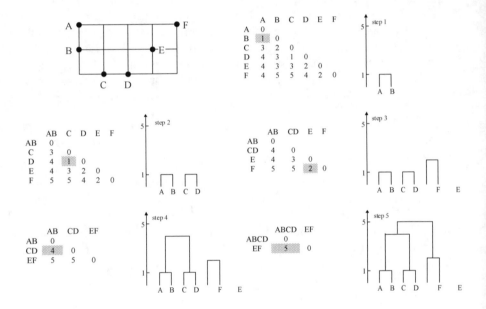

FIGURE 4.5
Stages of construction of a hierarchical tree from six individuals belonging to
a plane.

4.3.2   Hierarchy and Partitions

The points where the branches corresponding to the elements being grouped coincide are known as nodes. They are sometimes also known as "forks" when describing downward movement in the tree. The individuals to be classified are referred as leaf nodes. With I individuals, there are $(I-1)$ nodes, often numbered from $I+1$ to $2 \times I - 1$ (see Figure 4.6) according to the order of construction (the first I numbers are reserved for the leaf nodes; however, in some software, the leaf nodes are not numbered). In tracing a horizontal line to a given index, a partition is defined (the tree is said to be cut). In Figure 4.6, the cut at A defines a partition into two clusters $\{1,2,3,4\}$ and $\{5,6,7,8\}$; the cut at B defines a more precise partition into four clusters $\{1,2\}$, $\{3,4\}$, $\{5,6\}$ and $\{7,8\}$. By construction, these partitions are nested: each B-level cluster is included in the same cluster at level A.

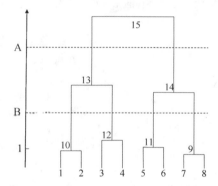

FIGURE 4.6
Hierarchy and partition.

Therefore, each hierarchical tree can be considered a sequence of nested partitions from the most precise (in which each individual is a class) to the most general (in which there is only one class).

4.4   Ward's Method

The principle of Ward's method is outlined above. It is applied to individuals situated in a Euclidean space. This is the most common case as it is that of a set of individuals described by a set of variables. When the data are quantitative (or categorical, respectively), we study cloud N_I evolving in $\mathbb{R}^K$ defined in Section 1.3.1 (and Section 3.4.2, respectively). This agglomerative method consists, at each stage of the process, of regrouping two elements (isolated or preclassed individuals) by maximizing the quality of the obtained partition.

4.4.1 Partition Quality

A partition can be said to be of high quality when

- The individuals within a cluster are homogeneous (small within-cluster variability).

- The individuals differ from one cluster to the next (high within-cluster variability).

If the individuals are within a Euclidean space, Huygens' theorem provides a framework for the analysis which is well suited to studying partitions. This theorem decomposes the total inertia (of the cloud of individuals) into two parts:

- The within-cluster inertia, based on the deviation between each point and the centre of gravity of the cluster to which it belongs.

- The between-clusters inertia, based on the deviation between each centre of gravity for a specific cluster and the overall centre of gravity.

Generally, it is expressed as

Total inertia = Between-clusters inertia + Within-cluster inertia.

If the individuals are described by just one quantitative variable (denoted y), we are left with the equation for a one-way analysis of variance. With I individuals (of the same weight, 1) divided into Q clusters, we denote y_{iq} the value for y of individual i in cluster q, $\bar{y}_q$ the average of y for the individuals in cluster q, I_q the number of individuals in cluster q, $\bar{y}$ the overall average of y. Huygens' theorem is expressed as

$$\sum_{q=1}^{Q}\sum_{i=1}^{I_q}(y_{iq} - \bar{y})^2 = \sum_{q=1}^{Q}I_q(\bar{y}_q - \bar{y})^2 + \sum_{q=1}^{Q}\sum_{i=1}^{I_q}(y_{iq} - \bar{y}_q)^2.$$

Generally speaking, there are K quantitative variables and the cloud of individuals evolves in $\mathbb{R}^K$ (see cloud N_I in PCA Section 1.3.1; Section 4.7.1 will show how to get back to this case if the variables are categorical). As the dimensions of $\mathbb{R}^K$ are orthogonal, Huygens' theorem is obtained by adding the inertias along each dimension. Thus, by denoting y_{iqk} the value for variable k of individual i in cluster q,

$$\sum_{k=1}^{K}\sum_{q=1}^{Q}\sum_{i=1}^{I_q}(y_{iqk} - \bar{y}_k)^2 = \sum_{k=1}^{K}\sum_{q=1}^{Q}I_q(\bar{y}_{qk} - \bar{y}_k)^2 + \sum_{k=1}^{K}\sum_{q=1}^{Q}\sum_{i=1}^{I_q}(y_{iqk} - \bar{y}_{qk})^2$$

$$\begin{array}{ccccc} \text{Total} & = & \text{Between-clusters} & + & \text{Within-cluster} \\ \text{inertia} & = & \text{inertia} & + & \text{inertia.} \end{array}$$

If we use this decomposition as a framework for analysis (i.e., if we use inertia to measure variability), then, when checking for partition quality, the same is achieved by minimising variability within-cluster or maximising variability between-clusters, as total variability is defined by the data. This is a great advantage for the user, who would find it extremely difficult to choose one of these two criteria. As a result, partition quality can be measured by

$$\frac{\text{Between-clusters inertia}}{\text{Total inertia}}.$$

This ratio indicates what can be imputed to the partition in terms of total variability. It is often multiplied by 100 in order to express it as a percentage. In one-dimensional cases, it coincides with the (square of) correlation ratio. For the data in Figure 4.5, by using the usual Euclidean distance and considering the partition into three clusters $\{A, B\}$, $\{C, D\}$, and $\{E, F\}$, this ratio is worth 0.8846. This partition thus expresses 88.46% of the individuals' variability. In other words, rather than considering the set of six individuals, we simply consider the three clusters; 88.46% of the variability of the data is nonetheless represented. When interpreting this percentage, the number of individuals and the number of clusters have to be taken into consideration. Indeed, by increasing the number of clusters, we can find partitions with extremely high percentages. Partitions in which each individual constitutes a cluster can even be said to have a percentage of 100%, although such a partition would be of no practical use. In our limited example, we can see that partitioning six individuals into three clusters, which halves the complexity of the data whilst still expressing 88.46% of it, is more than satisfactory.

4.4.2 Agglomeration According to Inertia

At step n of the agglomerative algorithm, the individuals are distributed in Q $(= I - n + 1)$ clusters obtained from previous steps. The difficulty is in choosing the two clusters (among Q) to be agglomerated. When grouping two clusters together, we move from a partition in Q clusters to a partition in $Q-1$ clusters; the within-cluster inertia can only increase (a direct consequence of Huygens' theorem). Grouping according to inertia means choosing the two clusters to be agglomerated so as to minimise the increase of within-cluster inertia. According to Huygens' theorem, combining two clusters this way leads to a decrease in between-clusters inertia, a decrease which is minimised.

Let us consider clusters p (with the centre of gravity g_p and sample size I_p) and q (with the centre of gravity g_q and sample size I_q). The increase $\Delta(p, q)$ in within-cluster inertia caused by grouping together clusters p and q is expressed as

$$\Delta(p, q) = \frac{I_p I_q}{I_p + I_q} d^2(g_p, g_q).$$

Choosing p and q in order to minimise $\Delta(p, q)$ means choosing

- Clusters whose centres of gravity are close together ($d^2(g_p, g_q)$ small).

- Clusters with small sample sizes ($\frac{I_p I_q}{I_p + I_q}$ small).

The first of these properties is instinctive. The second is less so but has an interesting consequence: agglomeration according to inertia tends to yield appropriate trees in the sense that the partitions are made up of clusters with samples of similar sizes. By applying this algorithm to the data in Figure 4.5, we obtain the tree in Figure 4.7; the level indices and the details of how they are calculated are summarised in Table 4.1.

The overall shape of the tree is identical to that obtained in Figure 4.5, with a different distance and agglomerative criterion: when a structure is strong, it is emphasised (almost) whatever the method selected. The major difference between the two hierarchies lies in level variability: agglomeration by inertia exacerbates the differences between the higher levels on the one hand, and the lower levels on the other, and this is especially due to the coefficient $\frac{I_p I_q}{I_p + I_q}$, which increases (almost) "mechanically" between the first levels (which agglomerates the elements with small sample sizes) and the last levels (which, in general, agglomerates clusters with large sample sizes).

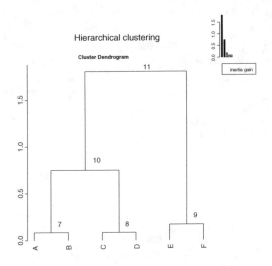

FIGURE 4.7

Tree obtained by applying Ward's algorithm to the data in Figure 4.5 and by using the usual Euclidean metric. Above right: bar chart of level indices, from the root node (node at the highest level) to leaf node (node at the lowest level). The numbers of the nodes have been added to the tree.

In certain software, it is possible to represent hierarchical trees which use the square root of the within-cluster inertia, increasing with a node's level index. As a result, the trees are more compact. In this work, we use the original criterion, that is, the within-cluster inertia increase.

TABLE 4.1
Indices Associated with Figure 4.7

Number of the node	p	q	$\frac{I_p I_q}{I_p+I_q}$	$d^2(g_p, g_q)$	Index	in %	Cumulative %	Within inertia	Within variance
7	2	1	0.5	0.167	0.083	2.88	100	0.083	0.250
8	4	3	0.5	0.167	0.083	2.88	97.12	0.083	0.250
9	6	5	0.5	0.333	0.167	5.77	94.23	0.167	0.500
10	8	7	1	0.750	0.750	25.96	88.46	0.917	1.375
11	9	10	1.33	1.354	1.806	62.50	62.50	2.889	2.889
Sum					2.889	100			

Note: The individuals are considered as nodes numbered in the order in which they appear in the file (here in alphabetical order).

4.4.3 Two Properties of the Agglomeration Criterion

1. In the representation of the tree, the quantity $\Delta(p,q)$ is used as the index. As might be expected, this index is ever-increasing (by denoting Δ_n the index associated with step n we obtain: $\Delta_n \geq \Delta_{n-1}$): first, we agglomerate the similar clusters and those with small sample sizes, and then go on to agglomerate more distant clusters and those with larger sample sizes. This first property is highly important: it guarantees that the tree has no inversions (an inversion occurs, for example, when element $\{c\}$ combines with the group $\{a, b\}$ at a lower level than that for the agglomeration between a and b (see Figure 4.8).

A   B   C

FIGURE 4.8
Example of a tree featuring an inversion.

2. The sum of all agglomeration criteria (from the same hierarchy) is equal to the total inertia for all of the individuals (in terms of their centre of gravity). Thus

$$\sum_{n=1}^{I-1} \Delta_n = \text{total inertia.}$$

This property is obtained easily by referring to the evolution of the partition of individuals at each stage of the tree's construction. At step 0, each individual represents a cluster and partition within-cluster inertia is nil. Throughout the algorithm, the number of clusters decreases and within-cluster inertia increases (from Δ_n to step n); at the end of the algorithm, all of the individuals are in the same cluster and within-cluster inertia is equal to total inertia. Thus an indexed hierarchy (obtained using this method) proposes a decomposition of the the total inertia (i.e., of the variability of the data) and fits into the overall approach to principal component methods. The difference is that this decomposition is conducted by clusters in one case and by components in the other.

4.4.4 Analysing Hierarchies, Choosing Partitions

Although hierarchies are constructed in an upward direction, they are generally analysed in a downward direction. Let us remember the aim of such hierarchies: to provide a visual representation of the variability of data or, from another point of view, of the overall similarities between individuals. In addition, the final node in a hierarchy must answer the question: If one had to summarise variability by dividing the individuals into two clusters, where would the partition be located? We must also highlight the fact that the term node evokes the idea of bringing two clusters together (the perspective of the upward construction) more than the subdivision into two clusters, hence the term "fork," which is sometimes used when approaching the tree in a downward manner.

With agglomeration according to inertia, the level of a node, when seen in a downward direction, quantifies that which is gained (in between-clusters inertia, or the decrease in within-cluster inertia) by separating the two clusters which it connects. In the example (see Figure 4.7 and Table 4.1), the separation into two groups expresses 62.50% of variability. If we consider the partition into three clusters, the separation created by node 10 (consider $\{a, b\}$ and $\{c, d\}$ rather than $\{a, b, c, d\}$) represents 25.96% of variability and thus yields a percentage of 62.50% + 25.96% = 88.46% for the three clusters partition.

Here, it would appear that a hierarchy is extremely useful for justifying the choice of a partition; this is indeed its main advantage in applications with anonymous individuals, such as surveys. Tangibly, we will account for

- The overall appearance of the tree; in the example in Figure 4.7, it evokes partitioning into three clusters.

- The levels of the nodes, to quantify the above; these levels can be represented by a bar chart visualising their decrease (graph represented in the top right-hand corner of Figure 4.7); each irregularity in this decrease evokes another division.

- The number of clusters, which must not be too high so as not to impede the concise nature of the approach.

- Cluster interpretability: even if it corresponds to a substantial increase in between-clusters inertia, we do not retain subdivisions that we do not know how to interpret: in much the same way, we retain those subdivisions which can be successfully interpreted, even if they correspond to greater increases in inertia. Luckily, in practice, dilemmas such as these are uncommon.

A visual examination of the hierarchical tree and the bar chart of level indices suggests a division into Q clusters when the increase of between-clusters inertia when passing from a $Q - 1$ to a Q clusters partition is much greater than that from a Q to a $Q + 1$ clusters partition. By using a downward approach (i.e., working from the largest partition), the following criterion is minimised:

$$\min_{q_{min} \leq q \leq q_{max}} \frac{\Delta(q)}{\Delta(q + 1)}$$

where $\Delta(q)$ is the between-clusters inertia increase when moving from $q - 1$ to q clusters, q_{min} (and q_{max}, respectively) the minimum (or maximum, respectively) number of clusters chosen by the user. The function **HCPC** (Hierarchical Clustering on Principal Components) implements this calculation after having constructed the hierarchy, and suggests an "optimal" level for division. When studying a tree, this level of division generally corresponds to that expected merely from looking at it. This level is thus most valuable when studying a great number of trees automatically.

4.5 Direct Search for Partitions: K-Means Algorithm

4.5.1 Data — Issues

The data are the same as for principal component methods: an individuals × variables table and a Euclidean distance. We consider that the variables are quantitative, without compromising generalisability as in the same way as for hierarchical clustering, Section 4.7.1 will show how to get back to this status when the variables are categorical. Partitioning algorithms approach hierarchical clustering primarily in terms of the following two questions.

- In practice, indexed hierarchies are often used as tools for obtaining partitions. In essence, would there not be a number of advantages in searching for a partition directly?

- When dealing with a great number of individuals, the calculation time required to construct an indexed hierarchy can become unacceptably long. Might we not achieve shorter calculation times with algorithms searching for a partition directly?

There are a number of partitioning algorithms: here, we shall limit our explanation to just one of them — K-means algorithm — which is in practice sufficient.

4.5.2　Principle

First, the number of clusters Q is determined. One might consider calculating all of the possible partitions and retaining only that which optimises a given criterion. In reality, the number of possible partitions is so high that the calculation time associated with this approach is unacceptably long when the number of individuals exceeds a certain number. We therefore use the iterative algorithm described below.

Let P_n be the partition of the individuals at step n of the algorithm and ρ_n the ratio [(between-clusters)/(total inertia)] of this partition P_n:

0. We consider a given initial partition P_0; we calculate ρ_0.

At step n of the algorithm,

1. We calculate the centre of gravity $g_n(q)$ for each cluster q of P_n.

2. We reassign each individual to the cluster q that it is closest to (in terms of distance to the centres of gravity $g_n(q)$); we obtain a new partition P_{n+1} for which we calculate the ratio ρ_{n+1}.

3. As long as $\rho_{n+1} - \rho_n >$ threshold (i.e., partition P_{n+1} is better than P_n) we return to phase 1. Otherwise, P_{n+1} is the partition we are looking for.

The convergence of this algorithm is ensured by the fact that, at each step, ρ_n decreases. In practice, this convergence is rather quick (generally less than five iterations even for a large amount of data). Figure 4.9 illustrates this algorithm on a dataset where individuals belong to plane.

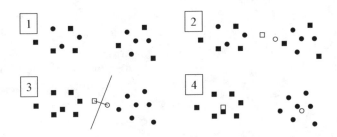

FIGURE 4.9
Illustration of the K-means algorithm for a simple case (the data present a number of well-distinct clusters corresponding to the number of clusters defined in the algorithm).

We are looking for a way to divide the 14 individuals into two clusters (a cluster of circles and a cluster of squares).

1. The individuals are randomly divided into two clusters.

2. The centres of gravity are calculated for each of the clusters (white circle and square).

3. Each individual is assigned to the cluster it is closest to (the perpendicular bisector of the segment linking the two centres of gravity is represented).

4. The centres of gravity for the new clusters are calculated.

If we reapply step three, nothing changes: the algorithm has converged.

4.5.3 Methodology

The above algorithm converges but not necessarily toward an overall optimum. In practice, the algorithm is conducted many times using different initial partitions P_0. The most satisfactory solution is retained. The partitions obtained by executing the algorithm a number of times can also be confronted. Sets of individuals which belong to the same cluster whatever the partition are referred to as "strong shapes." These strong shapes make up groups of individuals which are relatively stable compared to the initial partition: they highlight the denser areas (in space). However, this methodology also gives rise to some very small clusters, often of only one individual, made up of individuals situated between the high-density areas. Such individuals are empirically managed, the two main options being to assign them to the closest strong shape (with sufficiently large sample size) or to create a "residual" cluster grouping together all of the isolated individuals.

4.6 Partitioning and Hierarchical Clustering

When compared to hierarchical methods, partitioning strategies present two main advantages:

1. They optimise a criterion; in AHC, a criterion is optimised at each step, but we do not refer back to an optimal criterion concerning the tree itself.

2. They can deal with much greater numbers of individuals.

However, the groups need to be defined prior to the AHC. This is the origin of the idea of combining the two approaches to obtain a methodology that includes the advantages of each.

4.6.1   Consolidating Partitions

Following an AHC, an inspection of the hierarchy generally leads the user to focus on one partition in particular. This partition can be introduced as the initial partition of a partitioning algorithm. The partition resulting from this algorithm is finally conserved. In practice, the initial partition is never entirely replaced, but rather improved or "consolidated," the increase of the [(between inertia)/(total inertia)] ratio, although generally rather small, ensures that the clusters are (somewhat) more consistent and more appropriately separated. The minor inconvenience of this methodology is that the hierarchy produced by the AHC is not (exactly) consistent with the chosen partition.

4.6.2   Mixed Algorithm

When there are too many individuals to conduct an AHC directly, the following two-phase methodology can be implemented.

Phase 1. We partition into a high number of groups (100, for example). The partition obtained cannot be directly used for interpretation: there are a great number of groups and many of them are very close together. On the contrary, they are all homogeneous (low within-cluster inertia) and contain individuals which we are sure should not be separated.

Phase 2. We implement an AHC by taking the groups of individuals from phase one as elements to be classified (with each element's weight being the number, or rather the sum of the weights, of the individuals which it represents). In this way we obtain a hierarchy, which is, roughly, the top of the hierarchy that would be obtained by classifying the individuals themselves.

Another variation of phase 1 is to partition a number of times and to conserve the strong shapes for phase 2.

4.7   Clustering and Principal Component Methods

We have already mentioned this point: automatic clustering and principal component methods use similar approaches (exploratory analysis of a same data table) and differ in terms of representation methods (Euclidean clouds, indexed hierarchy or partition). This is indeed the origin of the idea of combining the two approaches to obtain an even richer methodology, an essential quality in exploratory statistics, since having many different perspectives can only reinforce the reliability of our conclusions and means we can choose that which is the most suitable for a given user (partitions are complex tools, but can be used by those with no statistical background). In this case, we use the same (Euclidean) distance between individuals for each method. First, because the choice of a distance must be made prior to the analysis as it is a sign of the assumptions that we might have about the similarities between individuals. Second because, if we want to study the influence of the choice of distance, it is better to do so using the same analysis method so as to avoid misleading comparisons.

4.7.1 Principal Component Methods Prior to AHC

Let us consider table X (of $I \times K$ dimensions) in which we wish to classify the rows. We perform a principal component method on X (PCA, CA, or MCA depending on the table) and retain all of the principal components (= coordinated of the rows on the component) with non-null variance (let S denote the number of such components and F_s the component of rank s). We bring these components together to construct table F (of $I \times S$ dimension). Tables X and F are equivalent in that they define the same distances between individuals. Among other things, the distance derived from the coordinates included in F is the usual Euclidean distance, and that even if the distance between the rows in X is not the usual Euclidean distance (for example, that of χ^2 when table X is analysed with a CA). In reality, the vectors u_s (associated with F_s) used to represent the individuals are an orthonormal basis, which is why the principal component maps obtained from a CA can be interpreted using the usual Euclidean distance even if the initial row space distance is a χ^2 distance.

As a result, the situation for programmers is facilitated as they can simply write one clustering programme with an individuals × quantitative variables table, and the usual distance between individuals as input. Different types of data can be accounted for by preprocessing the data using the most suitable principal component method (CA for a contingency table or Burt table; MCA for a categorical variables × individuals table). Using the two approaches together provides another new possible methodology: for AHC, we could simply retain only some of the principal components. The following two arguments are given for this:

- Eliminate the only dimensions which we are (practically) sure are only "noise," that is to say, the last. The components responsible for a very high percentage of the inertia (say 80% or 90%) are therefore retained; thus, the obtained hierarchy is considered to be more stable and clearer.

- Retain only those components which we know how to interpret; the main use of the hierarchy obtained is therefore simply to help in interpreting the results of the principal component method.

4.7.2 Simultaneous Analysis of a Principal Component Map and Hierarchy

This analysis mainly means representing the highest nodes of the hierarchy on the principal component map as centres of gravity for the individuals they group together. If we choose a partition, we limit our analysis to the centres of gravity of this sole partition. In a representation such as this, the two approaches complement one another in two ways:

1. First, we have a continuous view (the tendencies identified by the principal components) and a discontinuous view (the clusters obtained by the clustering) of the same set of data, all in a unique framework.

2. Second, the principal component map provides no information about the position of the points in the other dimensions; the clusters, defined from all of the dimensions, offer some information "outside of the plane;" two points close together on the plane can be in the same cluster (and therefore not too far from one another along the other dimensions) or in two different clusters (since they are far apart along other dimensions).

4.8 Clustering and Missing Data

Clustering programmes can easily be changed in order to apply them to incomplete data tables. Indeed, the paragraph § 1.8.4 (resp. § 3.7.4) shows that any quantitative table (or categorical table, respectively) containing missing data, can be completed using PCA (or MCA, respectively). The suitable principal component analysis, PCA or MCA depending on the nature of the data, can then be used to transform the completed table into a table bringing together S dimensions of the principal component analysis (see § 4.7.1). The standard clustering programme will be applied using these S quantitative dimenions.

4.9 Example: The Temperature Dataset

4.9.1 Data Description — Issues

In this section, we will be using the dataset regarding European capitals presented in the chapter on PCA (see Section 1.10). The objective here is to group the capitals together into comprehensive clusters so that the cities in a given cluster all present similar temperatures all year round. Once the clusters have been defined, it is important to describe them using the variables or specific individuals. To determine the number of clusters in which to group the capitals, we first construct an ascending hierarchical clustering.

4.9.2   Analysis Parameters

Clustering requires us to choose an agglomeration criterion (here we use Ward's criterion) along with a distance between individuals. Euclidean distance is suitable but it is also necessary to define whether or not the variables need to be standardised. We are again faced with the discussion we had in the PCA chapter (see Section 1.10.2.2) and we choose to work with standardised data. Furthermore, the distances between the capitals are defined using only 12 variables of monthly temperature, that is to say, from the active variables of the PCA.

Remark
The supplementary individuals (in the example, those towns which are not capital cities) are not used to calculate the distances between individuals and are therefore not used in the analysis.

The first two components of the PCA performed on the cities express over 98% of the information. All of the dimensions can therefore be retained as it does not affect clustering and can be used to decompose the total inertia of the PCA.

4.9.3   Implementation of the Analysis

The following lines of code first import the data and perform a PCA by specifying that all the components are retained using the argument `ncp=Inf` (`Inf` for infinite). Then an agglomerative hierarchical clustering is performed from the object `res.pca` containing the results of the PCA.

```
> library(FactoMineR)
> temperature <- read.table("http://factominer.free.fr/bookV2/temperature.csv",
    header=TRUE,sep=";",dec=".",row.names=1)
> res.pca <- PCA(temperature[1:23,],scale.unit=TRUE,ncp=Inf,
    graph=FALSE,quanti.sup=13:16,quali.sup=17)
> res.hcpc <- HCPC(res.pca)
```

Remark
It should be noted that it is possible to perform an agglomerative hierarchical clustering on a raw dataset by performing a nonstandardised PCA (using the argument `scale.unit=FALSE`) and retaining all of the components using `ncp=Inf`. This is the option by default when the **HCPC** function is performed directly on a dataset (a data frame).

The shape of the dendrogram (see Figure 4.10) suggests partitioning the capitals into three clusters. The optimal level of division calculated using the function **HCPC** also suggests three clusters. For example, in the first class, we find the coldest capitals (those with the weakest coordinates on the first axis of the principal component analysis). As indicated in Section 4.1 and represented in Figure 4.1, it is possible to switch the branches of each node in the tree so as to arrange the individuals according to the first principal component as far as possible. This is done using the default argument `order=TRUE`. If we want to sort the individuals according to another criterion, the individuals must first be sorted according to that criterion in the dataset prior to conducting the PCA, and then classified using the argument `order=FALSE` in **HCPC**.

The object `call$t` contains the results of the agglomerative hierarchical clustering. Specifically,

- The output of the function **agnes** (clustering function of the package cluster) in `call$t$tree`.

- The number of "optimal" clusters calculated (`$call$t$nb.clust`): this number is determined between the minimum and maximum number of clusters defined by the user and such that the ratio `$call$t$quot` might be as small as possible.

- The within-cluster inertia (`$call$t$within`) of the partitioning into n clusters; for $n = 1$ cluster (most general partition into one class) within-cluster inertia is worth 12, for 2 clusters 5.237, and so on.

- The increase in between-clusters inertia when moving from n clusters to $n+1$ (`$call$t$inert.gain`); for 2 clusters (i.e., moving from 1 to 2 clusters) the increase in between-clusters inertia is worth 6.763, for 3 clusters (i.e., moving from 2 to 3 clusters) 2.356, and so on.

- The ratio between two successive within-cluster inertias (for example, $0.550 = 2.881/5.237$).

```
$call$t$nb.clust
[1] 3

$call$t$within
 [1] 12.000  5.237  2.881  2.119  1.524  1.232  0.960  0.799  0.643  0.493
[11]  0.371  0.255  0.202  0.153  0.118  0.087  0.065  0.048  0.036  0.024
[21]  0.014  0.007  0.000

$call$t$inert.gain
 [1] 6.763 2.356 0.762 0.596 0.291 0.272 0.161 0.155 0.151 0.122 0.115 0.054
[13] 0.049 0.034 0.031 0.022 0.017 0.012 0.012 0.010 0.007 0.007

$call$t$quot
[1] 0.550 0.736 0.719 0.809 0.779 0.832 0.806 0.766
```

In order to draw the entire tree in three dimensions on the first principal component map (see Figure 4.11), we use the argument `t.levels="all"`:

```
> res.hcpc <- HCPC(res.pca,t.levels="all")
```

Defining the Clusters

The clusters are then defined and the results found in object `desc.var`. All of the variables from the initial dataset are used, whether the variables are

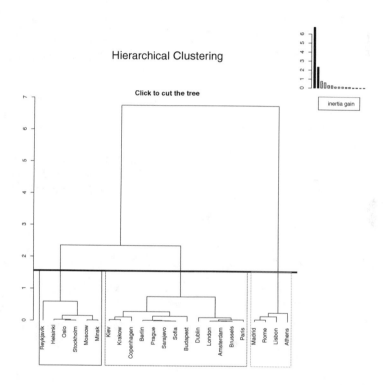

FIGURE 4.10
Temperature data: hierarchical tree.

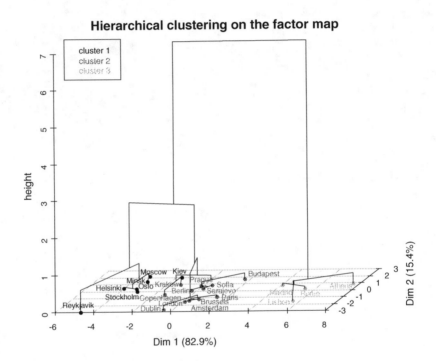

FIGURE 4.11
Temperature data: three-dimensional dendrogram on the first principal component map.

quantitative, categorical, active, or supplementary. In this case, the function yields the same results as the function **catdes** (see Section 3.7.2). These results are brought together in Table 4.2. The capitals from cluster 1 are characterised by below average temperatures throughout the year, and particularly in March (–1.14 degrees on average for the capitals in this cluster compared with 4.06 degrees for all of the capitals), October and February. In these cities, the latitude and thermal amplitude are all above average. None of the variables characterise the cities in cluster 2. The capitals in cluster 3 are typical as the average annual temperature (15.7 degrees) is much higher than the average for the other capitals (9.37 degrees). This cluster is characterised by the category *south* of the categorical variable *Area*: there are more southerly cities in this cluster than in the others. Indeed, 80% of southerly cities belong to cluster 3, and 100% of the cities in cluster 3 are southerly cities. These percentages are high because 21.7% of the cities are southerly.

These clusters can also be described by the principal components. To do so, a description identical to that carried out by the quantitative variables is conducted from the individuals' coordinates on the principal components. Table 4.3 thus shows that the capitals in cluster 1 (and 3, respectively) have a significantly weaker (or stronger, respectively) coordinate than the others in the first dimension. The coordinates in the third dimension are weaker for the capitals in cluster 2. It must be noted that only 1% of inertia is explained by component 3, and we shall therefore comment no further on this result.

It may be interesting to illustrate the cluster by using individuals specific to that class. To do so, two different kinds of specific individuals are suggested: paragons, that is to say, the individuals which are closest to the centre of the class, and the specific individuals, that is to say, those furthest from the centres of other clusters. To do so, the object `desc.ind$para` contains the individuals sorted into clusters and the distance between each individual and the centre of its class. The object `desc.ind$spec` contains the individuals sorted by cluster and the distance between each individual and the closest cluster centre (see Table 4.4). Thus, Oslo is the capital which best represents the cities in cluster 1, whereas Berlin and Rome are the paragons of clusters 2 and 3, respectively. Reykjavik is specific to cluster 1 because it is the city furthest from the centres of clusters 2 and 3, so we can consider it to be the most specific to cluster 1. Paris and Athens are specific to clusters 2 and 3.

4.10 Example: The Tea Dataset

4.10.1 Data Description — Issues

Let us return to the tea-drinking data presented in Chapter 3 on MCA in Section 3.1. The objective here is to propose a clustering of the 300 tea

TABLE 4.2
Temperature Data: **catdes** Outputs (See Section 3.7.2.3) Applied to the Partition into Three Clusters

```
> res.hcpc$desc.var
$test.chi2
        &p.value  df
Area     0.0012   6

$category
$category$'1'
NULL
$category$'2'
NULL
$category$'3'
            Cla/Mod Mod/Cla Global p.value v.test
Area=South      80     100 21.739   0.001  3.256

$quanti
$quanti$'1'
$'1'
```

	v.test	Mean in category	Overall mean	sd in category	Overall sd	p.value
Latitude	2.78	56.10	49.900	5.850	6.98	0.00549
Amplitude	2.14	22.00	18.800	4.840	4.61	0.03220
July	-1.99	16.80	18.900	2.450	3.33	0.04610
June	-2.06	14.70	16.800	2.520	3.07	0.03960
August	-2.48	15.50	18.300	2.260	3.53	0.01310
May	-2.55	10.80	13.300	2.430	2.96	0.01080
September	-3.14	11.00	14.700	1.670	3.68	0.00171
January	-3.26	-5.14	0.174	2.630	5.07	0.00113
December	-3.27	-2.91	1.840	1.830	4.52	0.00108
November	-3.36	0.60	5.080	0.940	4.14	0.00078
Annual	-3.37	5.50	9.370	0.767	3.56	0.00074
April	-3.39	4.67	8.380	1.550	3.40	0.00071
February	-3.44	-4.60	0.957	2.340	5.01	0.00058
October	-3.45	5.76	10.100	0.919	3.87	0.00055
March	-3.68	-1.14	4.060	1.100	4.39	0.00024

```
$quanti$'2'
NULL
$quanti$'3'
```

	v.test	Mean in category	Overall mean	sd in category	Overall sd	p.value
Annual	3.85	15.80	9.370	1.39	3.56	0.00012
September	3.81	21.20	14.700	1.54	3.68	0.00014
October	3.72	16.80	10.100	1.91	3.87	0.00020
August	3.71	24.40	18.300	1.88	3.53	0.00021
November	3.69	12.20	5.080	2.26	4.14	0.00022
July	3.60	24.50	18.900	2.09	3.33	0.00031
April	3.53	13.90	8.380	1.18	3.40	0.00041
March	3.45	11.10	4.060	1.27	4.39	0.00056
February	3.43	8.95	0.957	1.74	5.01	0.00059
June	3.39	21.60	16.800	1.86	3.07	0.00070
December	3.39	8.95	1.840	2.34	4.52	0.00071
January	3.29	7.92	0.174	2.08	5.07	0.00099
May	3.18	17.60	13.300	1.55	2.96	0.00146
Latitude	-3.23	39.40	49.900	1.52	6.98	0.00126

TABLE 4.3
Definition of Clusters (See Section 3.7.2.2) Resulting from the Clustering by Principal Components

```
> res.hcpc$desc.axe
$quanti
$quanti$'1'
       v.test Mean in category Overall mean sd in category Overall sd  p.value
Dim.1  -3.32            -3.37     1.69e-16              0.849       3.15 0.0009087
$quanti$'2'
       v.test Mean in category Overall mean sd in category Overall sd  p.value
Dim.3  -2.41           -0.175    -4.05e-16              0.218      0.355 0.0157738
$quanti$'3'
       v.test Mean in category Overall mean sd in category Overall sd  p.value
Dim.1   3.86             5.66     1.69e-16               1.26       3.15 0.00011196
```

drinkers into a few clusters corresponding to distinct tea-drinking profiles. For MCA, only the 19 questions relating to the way in which the participants drink tea were used as active variables; here again, these variables alone will be used to construct the clusters.

4.10.2   Constructing the AHC

Since the variables are categorical, the MCA performed prior to the clustering means the principal component coordinates can be used as quantitative variables. The last components of the MCA are generally considered as noise that should be disregarded in order to construct a more stable clustering. The first components are thus retained. Here, we choose 20 components which summarize 87% of the total inertia (we would rather keep more components than get rid of informative ones). After the MCA, an agglomerative hierarchical clustering is performed:

```
> library(FactoMineR)
> tea <- read.table("http://factominer.free.fr/bookV2/tea.csv",header=T,sep=";")
> res.mca<-MCA(tea,ncp=20,quanti.sup=22,quali.sup=c(19:21,23:36),graph=F)
> res.hcpc <- HCPC(res.mca)
```

The shape of the hierarchical tree, much like the bar chart of the inertias associated with the nodes, suggests partitioning into three clusters (see Figure 4.12).

We can then colour-code the individuals on the first two principal components map according to the cluster to which they belong (see Figure 4.13).

```
> plot(res.hcpc,choice="map",ind.names=FALSE)
```

The between-clusters inertia of partitioning into two clusters, 0.085 (first part of the following results), is less than the first eigenvalue of the MCA $\lambda_1 = 0.148$ (second part of the results below). This is always true and can be

TABLE 4.4

Paragons and Specific Individuals

```
> res.hcpc$desc.ind
$para
cluster: 1
     Oslo   Helsinki Stockholm    Minsk    Moscow
    0.339      0.884     0.922    0.965     1.770
------------------------------------------------
cluster: 2
   Berlin  Sarajevo  Brussels   Prague Amsterdam
    0.576     0.716     1.040    1.060     1.120
------------------------------------------------
cluster: 3
   Rome    Lisbon   Madrid   Athens
   0.36      1.74     1.84     2.17

$spec
cluster: 1
Reykjavik    Moscow   Helsinki    Minsk      Oslo
     5.47      4.34       4.28     3.74      3.48
------------------------------------------------
cluster: 2
   Paris  Budapest  Brussels   Dublin Amsterdam
    4.38      4.37      4.35     4.28      4.08
------------------------------------------------
cluster: 3
  Athens    Lisbon     Rome   Madrid
    7.67      5.66     5.35     4.22
```

explained by the fact that the principal component gives more nuances than partitioning into two clusters. Similarly, the map induced by the two main principal components expresses more inertia $(0.148 + 0.122 = 0.270)$ than partitioning into three clusters $(0.085 + 0.069 = 0.154)$. This is an advantage when we wish to summarise the information easily, for example, to plot the results. The MCA will be used to interpret the results in greater detail.

```
> round(res.hcpc$call$t$inert.gain,3)
 [1] 0.085 0.069 0.057 0.056 0.056 0.055 0.050
> round(res.mca$eig[,1],3)
 [1] 0.148 0.122 0.090 0.078 0.074 0.071 0.068
```

4.10.3   Defining the Clusters

To describe the characteristics of the individuals for each of the clusters, that is, their tea-drinking profiles, we define the clusters using the variables (object res.hcpc$desc.var, see Table 4.5) and the components (object res.hcpc$desc.axe, see Table 4.7). In this instance, it is less pertinent to describe the clusters from the individuals as they are unknown and thus

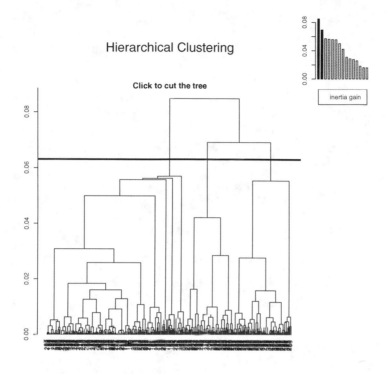

FIGURE 4.12
Tea data: hierarchical tree.

cannot be used for reference. Descriptions from categories (see Table 4.6) are simplified by retaining only the overexpressed categories associated with a p-value less than 2%.

The variables *place of purchase* and *format* best characterise the partitioning into three clusters (weakest probabilities equal to 8.47×10^{-79} and 3.14×10^{-47}, see Table 4.5).

More precisely, each of the clusters is characterised by one category of the variable *place of purchase* and one category of the variable *format*. The first cluster is characterised by individuals who buy tea in supermarkets in sachets: 85.9% of individuals who buy tea in supermarkets are in cluster 1 and 93.8% of the individuals in cluster 1 buy their tea in supermarkets. Similarly, cluster 2 is characterised by those who buy loose tea in specialist shops whereas cluster 3 is characterised by those who buy in both types of shop (supermarket and specialist), in both formats (sachet and loose). Other variables and categories are used to characterise each of the clusters although less distinctly (with a higher p-value).

The description of clusters from the principal components (see Table 4.7)

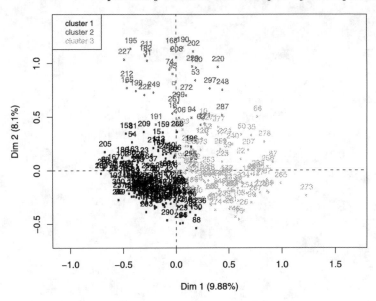

FIGURE 4.13
Tea data: representation of the partition on the principal component map.

shows that the individuals in cluster 1 have extremely weak coordinates on axes 1 and 2 (compared to individuals from other clusters). The individuals from cluster 2 have strong coordinates on component 2 and the individuals from cluster 3 have strong coordinates on component 1. We here retain the class-component pairs with a v-test greater than 3 as the components were used to construct the clusters.

4.11   Dividing Quantitative Variables into Classes

For certain analyses, it is preferable to transform a quantitative variable into a categorical variable. To do so, the variable must be divided into classes. The variable *age* for the tea data (see chapter on MCA) was noted as quantitative in the questionnaire. To highlight the nonlinear relationships with this variable, it must be recoded as categorical. Let us consider this variable *age* and study its transformation into a categorical variable.

   One strategy is to use "natural" classes defined prior to the analysis (for example, under 18 years, 18–30 years, etc.). Another strategy is to construct likely classes. In this case, we choose a number of classes, generally between four and seven so as to have enough classes but not too many:

TABLE 4.5

Tea Data: Description of the Partition into Three Clusters from the Variables

```
> res.hcpc$desc.var$test.chi2
                   p.value  df
place.of.purchase 8.47e-79  4
format            3.14e-47  4
type              1.86e-28 10
tearoom           9.62e-19  2
pub               8.54e-10  2
friends           6.14e-08  2
restaurant        3.54e-07  2
how               3.62e-06  6
variety           1.78e-03  4
sex               1.79e-03  2
frequency         1.97e-03  6
work              3.05e-03  2
afternoon.tea     3.68e-03  2
after.lunch       1.05e-02  2
after.dinner      2.23e-02  2
anytime           3.60e-02  2
sugar             3.69e-02  2
refined           4.08e-02  2
```

```
> tea <- read.table("http://factominer.free.fr/bookV2/tea.csv",header=T,sep=";")
> nb.clusters <- 4
> breaks <- quantile(tea[,22],seq(0,1,1/nb.clusters))
> Xqual <- cut(tea[,22],breaks,include.lowest=TRUE)
> summary(Xqual)
[15,23] (23,32] (32,48] (48,90]
     86      66      74      74
```

A third strategy is to choose the number of classes and their limits from the data. For example, we can use a histogram (see Figure 4.14) representing the distribution of the variable so as to define levels of division:

```
> hist(tea$age,col="grey",main="Histogram of the variable age",
    freq=FALSE,xlab="age",nclass=15)
```

In most cases, the choice of this division is not instantaneous as it is illustrated in Figure 4.14. It is convenient to use hierarchical clustering to determine a number of classes. Clusters can then be defined with the hierarchical clustering or by using, for example, the K-means method.

The following lines of code construct the hierarchical tree and consolidate the results using the K-means method (the K-means method converges very quickly when implemented on one variable alone):

```
> vari <- tea[,22]
> res.hcpc <- HCPC(vari,iter.max=10)
```

By default, the function **HCPC** constructs a hierarchical tree. The option

TABLE 4.6

Tea Data: Description of Partitioning into Three Clusters from the Categories

```
> res.hcpc$desc.var$category
$category$'1'
```

	Cla/Mod	Mod/Cla	Global	p.value	v.test
place.of.purchase=supermarket	85.9	93.8	64.0	4.11e-40	13.30
format=sachet	84.1	81.2	56.7	2.78e-25	10.40
tearoom=not.tearoom	70.7	97.2	80.7	2.09e-18	8.75
type=known.brand	83.2	44.9	31.7	2.78e-09	5.94
pub=not.pub	67.1	90.3	79.0	2.13e-08	5.60
friends=not.friends	76.9	45.5	34.7	3.42e-06	4.64
restaurant=not.restaurant	64.7	81.2	73.7	6.66e-04	3.40
type=shop.brand	90.5	10.8	7.0	2.40e-03	3.04
afternoon.tea=not.afternoon.t	67.9	50.6	43.7	5.69e-03	2.77
how=nothing.added	64.1	71.0	65.0	1.32e-02	2.48
work=not.work	63.4	76.7	71.0	1.41e-02	2.46
sugar=sugar	66.2	54.5	48.3	1.42e-02	2.45
anytime=not.anytime	64.0	71.6	65.7	1.45e-02	2.45
frequency=1 to 2/week	75.0	18.8	14.7	2.39e-02	2.26
frequency=1/day	68.4	36.9	31.7	2.61e-02	2.22
type=unknown	91.7	6.2	4.0	2.84e-02	2.19
age_Q=15-24	68.5	35.8	30.7	2.90e-02	2.18
after.lunch=not.after.lunch	61.3	89.2	85.3	3.76e-02	2.08
type=cheapest	100.0	4.0	2.3	4.55e-02	2.00

```
$category$'2'
```

	Cla/Mod	Mod/Cla	Global	p.value	v.test
place.of.purchase=specialist.shop	90.0	84.4	10.0	7.39e-30	11.40
format=loose	66.7	75.0	12.0	1.05e-19	9.08
type=luxury	49.1	81.2	17.7	4.67e-17	8.39
variety=green	27.3	28.1	11.0	7.30e-03	2.68
refined=refined	13.5	90.6	71.7	1.34e-02	2.47
sex=M	16.4	62.5	40.7	1.43e-02	2.45
restaurant=not.restaurant	13.1	90.6	73.7	2.59e-02	2.23
after.dinner=after.dinner	28.6	18.8	7.0	3.10e-02	2.16
exotic=not.exotic	14.6	71.9	52.7	3.23e-02	2.14

```
$category$'3'
```

	Cla/Mod	Mod/Cla	Global	p.value	v.test
place.of.purchase=supermarket+specialist	85.9	72.8	26.0	1.12e-33	12.10
format=sachet+loose	67.0	68.5	31.3	2.56e-19	8.99
tearoom=tearoom	77.6	48.9	19.3	2.35e-16	8.20
pub=pub	63.5	43.5	21.0	1.95e-09	6.00
friends=friends	41.8	89.1	65.3	2.50e-09	5.96
type=varies	51.8	63.0	37.3	2.63e-09	5.95
restaurant=restaurant	54.4	46.7	26.3	3.92e-07	5.07
how=other	100.0	9.8	3.0	3.62e-05	4.13
frequency=more than 2/day	41.7	57.6	42.3	6.13e-04	3.43
afternoon.tea=afternoon.tea	38.5	70.7	56.3	1.22e-03	3.23
work=work	44.8	42.4	29.0	1.32e-03	3.21
sex=F	37.1	71.7	59.3	4.90e-03	2.81
after.lunch=after.lunch	50.0	23.9	14.7	5.84e-03	2.76
how=lemon	51.5	18.5	11.0	1.32e-02	2.48
sugar=not.sugar	36.1	60.9	51.7	4.54e-02	2.00

Note: Output from the function **catdes** (see Section 3.7.2.3).

TABLE 4.7

Tea Data: Description of the Partition into Three Clusters from the Principal
Components

```
> res.hcpc$desc.axe
$quanti
$quanti$'1'
      v.test Mean in category Overall mean sd in category Overall sd  p.value
Dim.2  -7.80            -0.1320       4.93e-17           0.181      0.349 6.36e-15
Dim.1 -12.40            -0.2320      -2.00e-17           0.214      0.385 2.31e-35

$quanti$'2'
      v.test Mean in category Overall mean sd in category Overall sd  p.value
Dim.2  13.90             0.8120       4.93e-17           0.234      0.349 4.91e-44
Dim.4   4.35             0.2030      -3.35e-17           0.370      0.279 1.36e-05

$quanti$'3'
      v.test Mean in category Overall mean sd in category Overall sd  p.value
Dim.1  13.50             0.4520      -2.00e-17           0.252      0.385 1.89e-41
Dim.4  -4.73            -0.1150      -3.35e-17           0.292      0.279 2.30e-06
```

indicated here `iter.max=10` also conducts a K-means algorithm. The hierarchical tree (see Figure 4.15) suggests dividing the variable into four classes. This tree is constructed according to the values of the variable *age* on the abscissa axis.

We can then construct a new categorical variable `aaQuali`:

```
> max.cla <- unlist(by(res.hcpc$data.clust[,1],res.hcpc$data.clust[,2],max))
> aaQuali <- cut(vari,breaks = c(min(vari),max.cla),include.lowest=TRUE)
> summary(aaQuali)
[15,28] (28,42] (42,57] (57,90]
   130      68      64      38
```

This division seems of a better quality than the previous division into likely classes, as the hierarchical clustering detects the gaps in the distribution (see the bar chart in Figure 4.14).

If we wish to divide multiple quantitative variables into classes, it can be tedious to determine the number of classes to be chosen variable by variable from the hierarchical tree. In such cases, the function **HCPC** can be used, from which we can choose the optimal number of classes as determined by that function. The following lines of code are used to divide all of the quantitative variables from the dataset **don** into classes:

```
> don.quali <- don
> for (i in 1:ncol(don.quali)){
+    vari <- don.quali[,i]
+    res.hcpc <- HCPC(vari,nb.clust=-1,graph=FALSE)
+    maxi <- unlist(by(res.hcpc$data.clust[,1],res.hcpc$data.clust[,2],max))
+    aaQuali <- cut(vari,breaks = c(min(vari),maxi),include.lowest=TRUE)
+    don.quali[,i] <- aaQuali
+ }
```

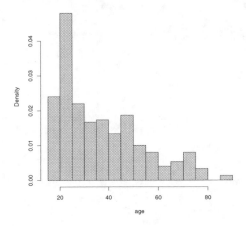

FIGURE 4.14
Tea data: bar chart for the variable *age*.

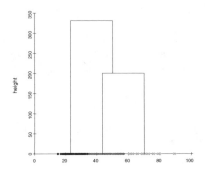

FIGURE 4.15
Tea data: dendrogram for the variable *age*.

The resultant table `don.quali` contains only categorical variables corresponding to the division into classes of each of the quantitative variables from the initial table `don`.

5

Visualisation

5.1 Data — Issues

As visualisation is at the heart of the principal component method, we believe that a separate chapter is required in order to fully illustrate what can be done with the FactoMineR package, and how it is used by other packages to view planes resulting from these principal component methods.

As we have already seen, the principle of the principal component method is that by changing perspective, it is possible to view a cloud of points in terms of its projected inertia from this new point of view. Once this new perspective has been identified, we must choose which information to view and how it should be represented.

This chapter is organised according to the principal component methods presented in this book, that is to say, principal component analysis (PCA), correspondence analysis (CA), and multiple correspondence analysis (MCA). For each of these methods, we analyse an ad hoc dataset with supplementary rows and columns. We explain how to graphically represent the numerical results stemming from the analysis according to its specific features.

Each paragraph describes the graphical representations specific to each principal component method. By default, the different clouds of points are represented on the first two dimensions of the different analyses. These representations can easily be obtained for other dimensions by changing the axes argument which, by default, is set at c(1,2).

5.2 Viewing Data Created by a Principal Component Analysis

As presented in Chapter 1, principal component analysis can be used to represent two clouds of objects: on the one hand, the cloud of individuals, and on the other, the cloud of variables. It is important to remember that aside from the active elements (individuals and quantitative variables), this method can be used to account for three kinds of illustrative (or supplementary) elements: individuals, quantitative variables, and qualitative variables.

5.2.1 Selecting a Subset of Objects — Cloud of Individuals

By default, the representation provided by the **plot.PCA** function, applied to an object such as the results of a PCA, is that of a cloud of individuals. When the analysis includes supplementary individuals, as well as supplementary qualitative variables, this representation includes these same supplementary elements, in which the qualitative variables are represented through their categories. Three types of objects are therefore represented: individuals, supplementary individuals, and the categories associated with the supplementary qualitative variables. The graph obtained using the following lines of code will be similar to the graph in Figure 1.21:

```
> library(FactoMineR)
> data(decathlon)
> res.pca <- PCA(decathlon, quanti.sup=11:12, quali.sup=13, ind.sup=c(36:41),
      graph=FALSE)
> plot.PCA(res.pca)
```

In order to choose to represent only one type of object, simply declare the other two types invisible. To do this, we use the argument `invisible` as follows:

```
> plot.PCA(res.pca, invisible=c("ind.sup","quali"))
```

By the same principle, if we want to represent the active and illustrative individuals, we use the argument `invisible` as follows:

```
> plot.PCA(res.pca, invisible="quali")
```

A more precise selection can be made in which it is possible to choose to represent just one subset of objects associated with a given type. The principle of this selection lies in the notion of representation quality, calculated for all the elements of the analysis (whether they are active or illustrative), as well as in the notion of contribution, calculated for the active elements alone. Once the representation dimensions are selected (by default these are dimensions 1 and 2), the FactoMineR package adds up, for these dimensions, the representation qualities (and contributions, respectively, here the raw contributions and not the relative contributions calculated for each dimension) in order to obtain a representation quality (and contribution, respectively) on the plane created by the dimensions.

It is therefore possible to choose to represent the 10 individuals which are best represented on the chosen plane (or those which contributed most to the construction of the plane, respectively), or indeed to represent individuals over a given threshold which we consider acceptable in terms of representation quality on the plane.

```
> plot.PCA(res.pca, select="cos2 10", invisible="quali")
> plot.PCA(res.pca, select="cos2 0.5", invisible="quali")
```

In the first part of this example, we chose to represent the 10 active individuals and the 10 illustrative individuals which are the best represented on the plane created by dimensions 1 and 2. In the second part, we chose to represent individuals for which the representation quality on the plane is higher than 0.5. In the first case, the user attributes the argument with a whole value between 1 and the number of individuals in the given dataset; in the second case, this value is between 0 and 1, with this latter value indicating that the object is perfectly represented on the plane in question.

It is also possible to represent a selection of individuals from their identity:

```
> plot.PCA(res.pca, select=c("Casarsa","Clay","Sebrle"), invisible="quali")
```

Whatever the means of selection, it is possible to change the way in which these unselected elements are represented using the argument unselect:

```
> plot.PCA(res.pca,select="cos2 10", invisible=c("ind.sup","quali"))
> plot.PCA(res.pca,select="cos2 10", invis=c("ind.sup","quali"), unselect=0)
> plot.PCA(res.pca,select="cos2 10", invis=c("ind.sup","quali"), unselect=1)
> plot.PCA(res.pca,select="cos2 10", invis=c("ind.sup","quali"), unselect=0.5)
> plot.PCA(res.pca,select="cos2 10", invis=c("ind.sup","quali"),
      unselect="yellow")
```

The first line of code corresponds to the default option: the unselected elements appear with a degree of transparency. The second line of code does not change the appearance of the unselected individuals: they appear in black, like the selected elements. The third line of code indicates that the unselected elements disappear. The fourth line of code is an intermediary between the second and third lines: the unselected elements appear in grey. To summarise, the argument unselect varies between 0 and 1, according to how we want the unselected elements to appear: they vary from black to white, passing through lighter and lighter shades of grey. Finally, the fifth line of code shows how to attribute unselected elements with a colour, which in this example is yellow.

5.2.2 Selecting a Subset of Objects — Cloud of Variables

To represent a cloud of variables in PCA, simply use the argument choix as follows:

```
> plot.PCA(res.pca, choix="var")
```

Just as for the cloud of individuals, by default this graph includes all of the information relating to quantitative variables, be they active or illustrative. It is therefore also possible to choose to represent only one type of variable, active or illustrative. It is also possible to create a graph based on a more specific selection, according to the same principle as explained in the previous section.

```
> plot.PCA(res.pca, choix="var", invisible="quanti.sup")
> plot.PCA(res.pca, choix="var", select="cos2 5")
```

In the second part of this example, we chose to represent the 5 active variables and the 5 illustrative variables which are the best represented on the plane created by dimensions 1 and 2.

It is possible to represent just some of the quantitative variables by specifying their names.

```
> plot.PCA(res.pca, choix="var", select=c("Long.jump","Rank","400m","100m"))
```

Just as for the representation of the cloud of individuals, it is possible to change the appearance of the unselected elements using the argument `unselect`:

```
> plot.PCA(res.pca, choix="var", invisible="quanti.sup",
      select=c("Long.jump","400m","100m"), unselect=0)
```

5.2.3   Adding Supplementary Information

Once the elements to be viewed have been chosen, it is important to represent them as best as possible by choosing appropriate colours, or by including supplementary qualitative information when appropriate/useful.

When representing the cloud of individuals, each type of object is represented by a colour code. By default, the individuals are black, the illustrative individuals are blue, and the categories associated with illustrative qualitative variables are pink. It is possible to change these colours using the arguments `col.ind`, `col.ind.sup`, and `quali` to change the colours of individuals, supplementary individuals and categories. The following line of code can be used to change the colour of supplementary objects:

```
> plot.PCA(res.pca, col.ind.sup="grey60", col.quali="orange")
```

Remark
When representing clouds of variables, each type of object is represented according to a colour code. By default, the active variables are black and the illustrative variables blue. These colours can be changed using the arguments `col.var`, `col.quanti.sup` as illustrated below:

```
> plot.PCA(res.pca, choix="var", col.var="orange", col.quanti.sup="grey")
```

Still with regard to representing the cloud of individuals, it is possible to attribute a colour code to the individuals according to a qualitative variable when this variable is one of the illustrative qualitative variables in the given dataset. For the qualitative variable in question, FactoMineR colours the individuals according to the category which characterises them. Using the argument `habillage`, the number or the name of this qualitative variable can be specified. In the same way, the following two lines of code can be used to colour individuals according to the categories of the variable Competition:

```
> plot.PCA(res.pca, habillage=13)
> plot.PCA(res.pca, habillage="Competition")
```

For this type of graph, it is possible to choose the colour of each of the categories, as illustrated in the following code, where the colour "orange" is attributed to the first category, and the colour "grey" to the second (the order of the categories can be obtained using the **levels** function):

```
> levels(decathlon[,13])
[1] "Decastar" "OlympicG"
> plot.PCA(res.pca, invisible=c("ind.sup","quali"), habillage=13,
    col.hab=c("orange","grey"))
```

In addition to identifying the association between individuals and categories using a colour code, it is possible to represent the variability of the individuals associated with a given category by drawing an ellipse around this category. To do this, we use the **plotellipses** function from the FactoMineR package and specify the qualitative variable(s) for which the ellipses are constructed using the argument keepvar:

```
> plotellipses(res.pca, keepvar=13)
```

We can use the same arguments as the function **plot.PCA** to obtain more precise representations. For example, to represent only those individuals with a representation quality of over 0.6 using the argument select = "cos2 0.6" (see graph in Figure 5.1):

```
> plotellipses(res.pca, keepvar=13, select="cos2 0.6")
```

5.3   Viewing Data from a Correspondence Analysis

As explained in Chapter 2, correspondence analysis (CA) can be used to analyse a contingency table from its constituent rows and columns. In order to do this, the CA identifies the dimensions which maximise the inertia of the cloud of rows (and columns, respectively) projected on these dimensions, and inertia is calculated from the distance to the Chi Square. In the analysis, the rows and columns take on a symmetric role. In this section, we shall see firstly how to select the rows (or columns), and secondly how to add supplementary information.

5.3.1   Selecting a Subset of Objects — Cloud of Rows or Columns

By default, the representation provided by the **plot.CA** function applied to an object such as CA results is that of a cloud of rows and columns. When

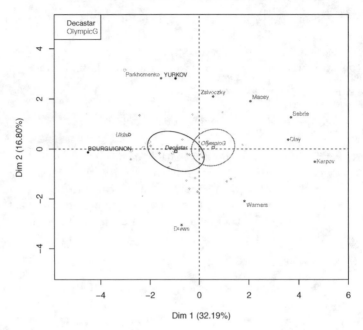

FIGURE 5.1
Graph showing a selection of individuals according to their representation quality and confidence ellipses around the categories.

the analysis contains supplementary elements (rows and/or columns), this representation includes these same supplementary elements. Four types of objects are therefore represented: active rows, active columns, supplementary rows, and finally supplementary columns.

```
> library(FactoMineR)
> data(children)
> res.ca <- CA(children, row.sup = 15:18, col.sup = 6:8, graph=FALSE)
> plot.CA(res.ca)
```

By the same principle as previously described, in order to represent only certain types of elements the other types must be masked using the argument `invisible` as follows:

```
> plot.CA(res.ca, invisible=c("col","row.sup","col.sup"))
> plot.CA(res.ca, invisible=c("row.sup","col.sup"))
```

In the first line of code, we choose to represent only the rows of the dataset; in the second line of code, we choose to represent only the active elements by masking the supplementary elements.

To refine the selection and to represent just one subset of rows and/or columns according to their representation quality (and contribution, respectively), simply use the arguments `selectRow` and `selectCol` as follows:

```
> plot.CA(res.ca, selectRow="cos2 3", selectCol="cos2 2")
> plot.CA(res.ca, selectRow="cos2 0.8", selectCol="cos2 0.8")
```

The first line of code is used to represent the 3 active lines and the 3 supplementary lines as well as the 2 active columns and the 2 supplementary columns for which the representation quality on the given plane is highest (see Fig. 5.2). The second line of code is used to select the elements for which the representation quality on the given plane is higher than 0.8.

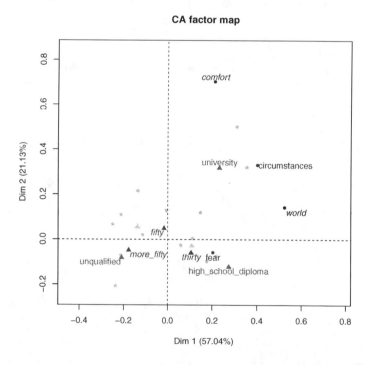

FIGURE 5.2

Graph with a selection based on representation quality.

Finally, to represent just one selection based on the identities of the rows or columns, we use the following code:

```
> plot.CA(res.ca, selectRow=c("work","hard"), selectCol=c("cep","university"))
```

Using the argument `unselect`, it is possible to change the appearance of the unselected elements (see § 5.2.1), as shown with the following code:

```
> plot.CA(res.ca, selectCol="cos2 5", selectRow="cos2 5", unselect=0)
> plot.CA(res.ca, selectCol="cos2 5", selectRow="cos2 5", unselect=1)
> plot.CA(res.ca, selectCol="cos2 5", selectRow="cos2 5", unselect="yellow")
```

5.3.2   Adding Supplementary Information

By definition, at the intersection of a row i and a column j of a contingency table, we find n_{ij}, the number of individuals associated both with the categories corresponding to the given row and column, respectively. If the table is composed of I rows and J columns, the data from the contingency table can be assimilated as the accomplishment of a random variable according to a multinomial distribution corresponding to the accomplishment of $I \times J$ events, each with a probability of being carried out of n_{ij}/n, where $n = \sum_{i,j} n_{ij}$. Contingency tables can be generated from this distribution and represented as supplementary elements in the correspondence analysis of the table initially studied. In this way, confidences ellipses can be drawn around the representation of categories from the CA of the data table. This is precisely the objective of the function **ellipseCA** which draws ellipses around categories in order to assess how their position varies on the factorial plane. This function is used as follows:

```
> ellipseCA(res.ca)
```

To represent just a subset of ellipses, the `ellipse` argument can be modified to select only the ellipses around the rows or around the columns. We can also use the arguments `col.row.ell` or `col.col.ell` and choose the colour `"transparent"` so that no ellipse is shown:

```
> ellipseCA(res.ca, ellipse="row", invisible=c("row.sup","col.sup"),
    col.row.ell=c(rep("grey",2),rep("transparent",12)))
```

In this example, we represent ellipses around active rows: the first two active rows have grey ellipses whereas those which follow have no ellipses.

5.4   Viewing Data from a Multiple Correspondence Analysis

According to Chapter 3, MCA can be seen as a CA applied to a complete disjunctive table. From this perspective, we can expect a representation of rows (statistical individuals) and columns (categories associated with qualitative variables) which is extremely cluttered, particularly as this method is well suited to statistical processing of questionnaire data, particularly for those with a large number of individuals described by a large number of qualitative variables, and consequently a great number of categories to represent.

Due to this complexity, we will analyse the superimposed representation of the rows and columns in two phases; firstly through the representation of the cloud of individuals, and secondly through the representation of the cloud of categories. We will complete this analysis with an analysis of variables, using a representation of supplementary quantitative variables like that provided by the PCA, using a representation of all of the variables thanks to which the qualitative information can be explained not at the level of the categories, but rather at the level of the variables themselves.

5.4.1 Selecting a Subset of Objects — Cloud of Individuals

By default, the **plot.MCA** function represents four types of objects: active individuals, supplementary individuals, active categories, and supplementary categories. To view the individuals alone, the argument `invisible` is used as follows:

```
> plot.MCA(res.mca, invisible=c("var","quali.sup"))
> plot.MCA(res.mca, invisible=c("var","quali.sup"), label="none")
> plot.MCA(res.mca, invisible=c("var","quali.sup"), label="ind.sup")
```

The second line of code is used to view the cloud of individuals without their identity, whereas in the third line of code it is possible to add the identity of some supplementary individuals to the analysis. This can be important when we have a particular interest in these individuals.

In the same way as for PCA or CA, it is possible to select individuals according to their representation quality or contribution (see § 5.2.1):

```
> plot.MCA(res.mca, invisible=c("var","quali.sup"), select="cos2 50")
> plot.MCA(res.mca, invisible=c("var","quali.sup"), select="cos2 0.5")
```

The first line of code is used to view the 50 active individuals (or supplementary, respectively), with the highest representation quality on the given plane; the second line of code is used to view those individuals with a representation quality higher than 0.5.

The argument `unselect` is used to visually reinforce this selection by accentuating the level of transparency for unselected individuals:

```
> plot.MCA(res.mca, invisible=c("var","quali.sup"), label="none",
      select="cos2 0.5", unselect=0.9)
```

5.4.2 Selecting a Subset of Objects — Cloud of Categories

The cloud of categories is at the heart of MCA analysis as it makes it possible to interpret the principal dimensions of variability of the cloud of individuals. Its analysis can be carried out in two phases: firstly, through the categories associated with active variables, then through the categories associated with supplementary variables. These two representations are obtained using the argument `invisible` as follows:

```
> plot.MCA(res.mca, invisible=c("ind","ind.sup","quali.sup"))
> plot.MCA(res.mca, invisible=c("ind","ind.sup","var"))
```

In much the same way as in the previous sections, we will view only the 10 active categories and the 10 illustrative values which are represented the best, along with the 15 categories which contributed to the construction of the plane (thus only active categories will be represented). This is done using the argument `selectMod`. In order to better distinguish between the selected and unselected elements, we reinforce the opacity of the mask by attributing a value near 1 to the argument `unselect`.

```
> plot.MCA(res.mca, invisible=c("ind","ind.sup"), selectMod="cos2 10",
      unselect=0.9)
> plot.MCA(res.mca, invisible=c("ind","ind.sup"), selectMod="contrib 15",
      unselect=0.9)
```

5.4.3　Selecting a Subset of Objects — Clouds of Variables

In MCA, a variable can be represented in two different ways. The first, which is relatively classic, is used to view the illustrative qualitative variables through a representation of the type of correlation circle. The second, which is less common, is used to represent the variables, whatever their nature (see Section 3.4.3). For a given pair of dimensions, the coordinates of a quantitative variable (or qualitative, respectively) in the perspective generated by the dimensions, are obtained by calculating the square of the correlation coefficient (of the correlation ratio) between the given quantitative variable and each of the dimensions.

```
> plot.MCA(res.mca, choix="quanti.sup", title="Quantitative variables graph")
> plot.MCA(res.mca, choix="var", title="Graph of the variables")
```

These representations can be used to interpret the dimensions generated by the MCA in terms of the variables (and not their categories, in the case of qualitative variables). It is therefore important for them to be perfectly useable. The following code can be used to select variables according to their type:

```
> plot.MCA(res.mca, choix="var", invisible=c("quali.sup","quanti"))
```

5.4.4　Adding Supplementary Information

It is important to remember that in MCA the cloud of individuals and the cloud of categories exist within the same space. This is reminiscent of PCA where individuals and supplementary qualitative variables, through their categories, are represented in the same graph. In MCA, supplementary information is therefore added according to the same principles as in PCA.

To change the colours of different types of objects in the representation of individuals, users can change the parameters `col.ind`, `col.ind.sup`, `col.var` (in MCA, the term variable refers to the notion of active qualitative variable) and `col.quali.sup`.

```
> plot.MCA(res.mca, col.var="black", col.ind="yellow", col.quali.sup="orange")
```

To draw confidence ellipses around categories, we use the function **plotellipses**, firstly on all the variables, and then to target a specific variable.

```
> plotellipses(res.mca)
> plotellipses(res.mca, keepvar="where")
```

5.5 Alternatives to the Graphics Function in the FactoMineR Package

There are alternatives to the FactoMineR package for providing graphical representations of factorial planes, but which are nonetheless based on numerical output. Here we will present two packages: Factoshiny and factoextra.

5.5.1 The Factoshiny Package

As its name suggests, the Factoshiny package uses *shiny* technology to produce a graphical interface of the FactoMineR package in the form of an *html* page (see Figure 5.3). Users can fill out different fields to fine tune their analyses. Once the fields are completed, the analyses are conducted instantly and provide the chosen graphs. Cursors enable users to change the extent to which objects are selected using representation quality or contribution and other cursors can be used to change the size of the font used to label the graph. The graphs are updated immediately, which makes these selections easier. This package makes interacting with R and FactoMineR simpler, thus facilitating selections and the addition of supplementary information as described in the previous sections. In addition to its FactoMineR interfacing capabilities and its ability to effortlessly create complex graphs, the Factoshiny package can also generate the code used to construct the graphs. The following code indicates how Factoshiny can be used to carry out a PCA.

```
> library(Factoshiny)
> data(decathlon)
> res.shiny <- PCAshiny(decathlon)
```

Then the interface is used to choose the parameters of the PCA analysis as well as the parameters of the graphical outputs. It is also possible to first perform the PCA analysis and then to use the PCA outputs in order to modify the graphs.

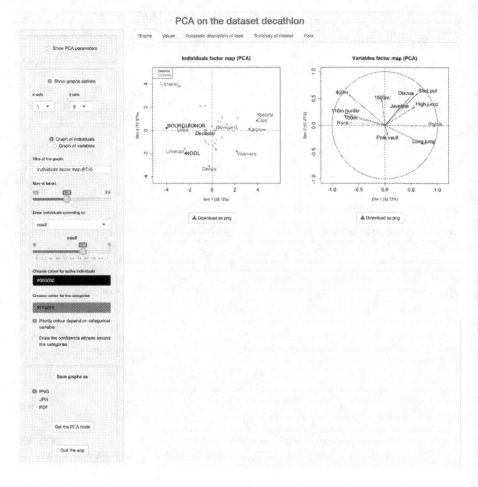

FIGURE 5.3
PCA with Factoshiny.

```
> res.pca <- PCA(decathlon, quanti.sup=11:12, quali.sup=13, ind.sup=c(36:41),
    graph=FALSE)
> res.shiny <- PCAshiny(res.pca)
```

It is also possible to reuse the object resulting from Factoshiny in order to further modify the graphs, using the configuration described previously:

```
> res.shiny2 <- PCAshiny(res.shiny)
```

The functions **CAshiny** and **MCAshiny** operate on exactly the same principle. From a certain point of view, the Factoshiny package is a natural evolution of the RcmdrPlugin.FactoMineR package: compared to its predecessor, Factoshiny includes the latest graphical updates in the FactoMineR package, it is more user friendly, and in particular, it enables the user to handle and explore data directly.

5.5.2   The factoextra Package

The factoextra package is an interesting alternative to the graphical functions offered by FactoMineR. It is based on the *Grammar of Graphics* concept put forward by Wilkinson, Anand and Grossman in 2005 and adapted and developed for R by Hadley Wickham using the ggplot2 package. The factoextra package can be used to construct graphs sequentially, gradually adding layers of graphical elements. The following code is used to obtain the default representation of individuals:

```
> library(factoextra)
> data(decathlon)
> res.pca <- PCA(decathlon, quanti.sup=11:12, quali.sup=13, ind.sup=c(36:41),
      graph=FALSE)
> fviz`pca`ind(res.pca)
```

To change the title of the graph or the legends associated with the dimensions, simply add a layer using the **labs** and write:

```
> library(factoextra)
> res.pca <- PCA(decathlon, quanti.sup=11:12, quali.sup=13, ind.sup=c(36:41),
      graph=FALSE)
> fviz`pca`ind(res.pca) + labs(title="PCA", x="PC1", y="PC2")
```

Below are three ways of obtaining a representation of individuals according to their contribution (see Figure 5.4):

```
> fviz`pca`ind(res.pca, col.ind="contrib")
> fviz`pca`ind(res.pca, col.ind="contrib") + scale`color`gradient2(low="blue",
      mid="white",high="red", midpoint=4)
> fviz`pca`ind(res.pca, col.ind="contrib") + scale`color`gradient2(low="blue",
      mid="white", high="red", midpoint=4) + theme`minimal()
```

As before, the first line of code is used to obtain the default representation of individuals according to their contribution, the second to change the colour gradient, and the third to declutter the background of the graph.

Documentation for the factoextra package is available at the www.sthda.com.

5.6   Improving Graphs Using Arguments Common to Many **FactoMineR** Graphical Functions

Many arguments which are common to all the graphical functions in FactoMineR can be used to configure the graphs. Here we will show how these

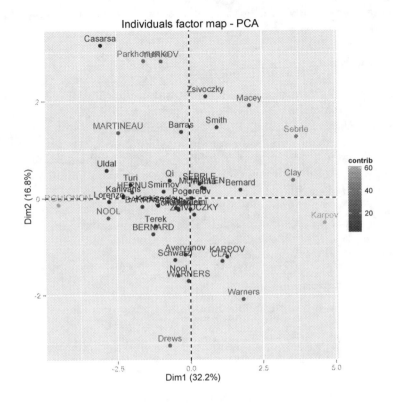

FIGURE 5.4
The factoextra Package: graph of individuals according to their contribution.

arguments are used. The following lines of code provide the graph in Figure 5.5.

```
> library(FactoMineR)
> data(decathlon)
> res.pca <- PCA(decathlon, quanti.sup=11:12, quali.sup=13, ind.sup=c(36:41),
      graph=FALSE)
> plot.PCA(res.pca, select="cos2 0.6", unselect=1, habillage="Competition",
      title="Graphe des individus", autoLab="yes", shadow=TRUE,
      cex=0.8, cex.main=1.2, cex.axis=0.8, cex.lab=0.8)
```

The arguments used in the last line of code require some explanation. We select those individuals with a representation quality higher than 0.6 on the plane (`select="cos2 0.6"`), we display the unselected elements without transparency (`unselect=1`), we colour the individuals according to the category they take for the Competition variable (`habillage="Competition"`), we change the title of the graph using the argument `title`, and then we position the objects' labels so as to create the least possible overlap using the argument `autoLab="yes"`. This argument makes use of an algorithm which aims to minimise overlap in the labelling of objects. The algorithm can take a long time to run when there are a lot of labels. This option can be disconnected using the argument `autoLab="no"`. By default, the option `autoLab="auto"` launches the algorithm when there are fewer than 50 objects (as if we had used `autoLab="yes"`), and otherwise does not launch it (as if we had used `autoLab="no"`). If the same instruction is launched repeatedly, the graphs can be slightly different as the position of the labels can change slightly according to the convergence of the algorithm. The argument `shadow=TRUE` can be used to add a slight shadow under the labels to improve their readability on the axes (for example, we can see that where the category Decastar appears, the horizontal axis is not drawn). Finally, the last arguments can be used to change the size of the points' fonts (`cex`), the title (`cex.main`), the axes (`cex.axis`), or the axis labels (`cex.lab`). For all of these arguments, a value of 1 represents the default value. Values lower than 1 lead to a smaller font, and values greater than 1 will create a larger font.

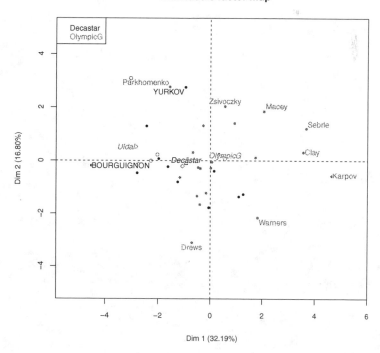

FIGURE 5.5
Graph illustrating different arguments common to many FactoMineR graphical functions.

Appendix

A.1   Percentage of Inertia Explained by the First Component or by the First Plane

Here, we aim to test the percentage of inertia explained by a component, and then the percentage of inertia expressed by the first plane. To do so, we simulated 10,000 datasets for a number of individuals I and a number of normally distributed independent variables K. We then conduct a standardised principal component analysis (PCA) (standardised variables) for each dataset and then calculate the percentage of inertia explained by one component and that expressed by one plane. In Tables A.1 and A.2 (respectively Tables A.3 and A.4) we define the quantile 0.95 of the 10,000 percentages of inertia of the first component (and the first plane, respectively) obtained for a given table size (I and K).

Thus, comparing the percentage of inertia of a component or plane with the associated value in the table amounts to testing the null hypothesis H_0: the percentage of inertia explained by the first component (and the first plane, respectively) is not significantly greater than that obtained with the (normally distributed) independent data.

TABLE A.1

95th percentile of the percentage of inertia explained by the first component of 10,000 PCAs performed on tables made up of independent variables (the number of individuals varies between 5 and 100, and the number of variables from 4 to 16). For example, for a table with $I = 30$ individuals and $K = 10$ variables, 95% of the percentages of inertia explained by the first component are less than 23.6%.

nbind	Number of variables												
	4	5	6	7	8	9	10	11	12	13	14	15	16
5	72.6	67.6	63.3	60.4	57.9	55.5	53.9	52.6	51.3	50.1	49.1	48.4	47.5
6	67.6	61.8	57.6	54.7	52.4	50.4	48.7	46.9	45.8	44.6	43.6	42.9	42.0
7	64.0	58.3	54.0	50.9	48.3	46.1	44.5	42.9	41.8	40.4	39.8	38.8	38.1
8	60.7	54.9	50.7	47.7	45.2	43.1	41.3	40.1	38.7	37.4	36.5	35.9	35.0
9	58.6	52.3	48.7	45.0	42.7	40.8	39.1	37.7	36.3	35.2	34.3	33.5	32.5
10	56.8	50.5	46.4	43.5	40.7	38.6	36.9	35.7	34.4	33.4	32.1	31.5	30.7
11	55.0	48.8	44.6	41.6	39.0	37.2	35.4	33.9	32.8	31.7	30.8	29.7	29.1
12	53.3	47.5	43.2	40.1	37.7	35.6	34.1	32.5	31.5	30.3	29.4	28.6	27.9
13	52.0	46.2	41.8	39.0	36.4	34.5	32.9	31.3	30.2	29.1	28.2	27.4	26.7
14	51.0	45.2	40.9	37.8	35.5	33.3	31.7	30.3	29.0	28.1	27.2	26.4	25.6
15	50.1	44.1	40.0	36.8	34.4	32.4	30.8	29.4	28.3	27.3	26.5	25.5	24.7
16	49.3	43.2	39.2	36.0	33.7	31.6	29.9	28.7	27.4	26.5	25.5	24.7	24.0
17	48.4	42.3	38.3	35.2	32.9	31.0	29.2	27.9	26.7	25.7	24.9	24.0	23.3
18	47.6	41.8	37.6	34.5	32.2	30.2	28.7	27.1	26.0	25.1	24.2	23.4	22.7
19	46.9	41.1	36.8	33.9	31.5	29.7	28.0	26.6	25.6	24.5	23.5	22.8	22.1
20	46.1	40.5	36.3	33.5	30.9	29.0	27.4	26.1	25.0	24.0	23.0	22.3	21.6
25	44.0	38.1	33.9	31.0	28.6	26.9	25.2	23.8	22.8	21.9	21.0	20.3	19.6
30	41.9	36.4	32.4	29.4	27.1	25.1	23.6	22.4	21.3	20.3	19.5	18.8	18.1
35	40.7	35.0	31.0	28.1	25.9	23.9	22.5	21.2	20.1	19.2	18.4	17.7	17.0
40	39.7	34.0	30.1	27.1	24.7	23.0	21.6	20.3	19.3	18.3	17.5	16.8	16.2
45	38.8	33.0	29.1	26.3	24.0	22.3	20.8	19.6	18.5	17.6	16.8	16.1	15.5
50	38.0	32.4	28.5	25.6	23.4	21.6	20.1	18.9	17.9	17.0	16.2	15.6	15.0
100	34.1	28.5	24.8	21.9	19.9	18.2	16.9	15.7	14.7	14.0	13.2	12.6	12.0

TABLE A.2

95th percentile of the percentage of inertia explained by the first component of 10,000 PCAs performed on tables made up of independent variables (the number of individuals varies between 5 and 100, and the number of variables from 17 to 200). For example, for a table with $I = 50$ individuals and $K = 30$ variables, 95% of the percentages of inertia explained by the first component are less than 10.4%.

nbind	\multicolumn{12}{c}{Number of variables}
	17	18	19	20	25	30	35	40	50	75	100	150	200
5	46.9	46.2	45.5	45.0	42.9	41.3	39.8	39.0	37.3	35.0	33.6	32.0	31.0
6	41.1	40.7	40.1	39.5	37.4	35.6	34.5	33.5	31.8	29.5	28.2	26.6	25.7
7	37.2	36.7	36.0	35.6	33.5	31.8	30.4	29.6	28.1	25.8	24.5	23.0	22.1
8	34.4	33.7	33.1	32.6	30.4	28.8	27.6	26.7	25.2	23.1	21.8	20.4	19.5
9	32.1	31.3	30.8	30.2	28.0	26.5	25.4	24.4	23.0	21.0	19.7	18.3	17.5
10	30.0	29.5	28.8	28.4	26.2	24.6	23.6	22.7	21.4	19.3	18.1	16.7	15.9
11	28.5	27.8	27.3	26.8	24.7	23.3	22.1	21.3	19.9	17.9	16.8	15.4	14.6
12	27.1	26.5	25.9	25.5	23.5	22.0	20.9	20.0	18.7	16.7	15.6	14.3	13.6
13	26.0	25.3	24.9	24.2	22.3	20.9	19.8	19.0	17.7	15.7	14.7	13.4	12.7
14	25.0	24.4	23.9	23.4	21.3	20.0	18.9	18.1	16.8	14.9	13.9	12.6	11.9
15	24.1	23.5	23.0	22.5	20.7	19.2	18.1	17.3	16.1	14.2	13.2	12.0	11.2
16	23.5	22.9	22.3	21.7	19.9	18.5	17.4	16.6	15.4	13.6	12.5	11.3	10.7
17	22.7	22.2	21.6	21.1	19.2	17.8	16.8	16.0	14.8	13.0	12.0	10.8	10.1
18	22.1	21.5	21.0	20.4	18.6	17.2	16.3	15.4	14.2	12.5	11.5	10.3	9.7
19	21.4	20.9	20.4	19.9	18.0	16.7	15.8	14.9	13.8	12.1	11.1	9.9	9.3
20	21.0	20.4	20.0	19.4	17.6	16.3	15.3	14.5	13.3	11.6	10.6	9.5	8.9
25	19.0	18.4	17.9	17.4	15.7	14.5	13.5	12.8	11.7	10.0	9.1	8.1	7.5
30	17.5	17.0	16.6	16.1	14.4	13.2	12.3	11.5	10.5	8.9	8.1	7.1	6.5
35	16.5	16.0	15.5	15.1	13.4	12.2	11.3	10.6	9.6	8.1	7.3	6.4	5.8
40	15.6	15.2	14.7	14.2	12.6	11.5	10.6	10.0	8.9	7.5	6.7	5.8	5.3
45	14.9	14.4	14.0	13.6	12.0	10.9	10.0	9.4	8.4	7.0	6.3	5.4	4.9
50	14.4	13.9	13.5	13.1	11.5	10.4	9.6	9.0	8.0	6.6	5.9	5.0	4.6
100	11.6	11.1	10.7	10.3	8.9	7.9	7.2	6.6	5.8	4.7	4.0	3.3	2.9

TABLE A.3
95th percentile of the percentage of inertia explained by the first component
of 10,000 PCAs performed on tables made up of independent variables (the
number of individuals varies between 5 and 100, and the number of
variables from 4 to 16). For example, for a table with $I = 30$ individuals
and $K = 10$ variables, 95% of the percentages of inertia explained by the
first plane are less than 41.1%.

nbind	Number of variables												
	4	5	6	7	8	9	10	11	12	13	14	15	16
5	96.5	93.1	90.2	87.6	85.5	83.4	81.9	80.7	79.4	78.1	77.4	76.6	75.5
6	93.3	88.6	84.8	81.5	79.1	76.9	75.1	73.2	72.2	70.8	69.8	68.7	68.0
7	90.5	84.9	80.9	77.4	74.4	72.0	70.1	68.3	67.0	65.3	64.3	63.2	62.2
8	88.1	82.3	77.2	73.8	70.7	68.2	66.1	64.0	62.8	61.2	60.0	59.0	58.0
9	86.1	79.5	74.8	70.7	67.4	65.1	62.9	61.1	59.4	57.9	56.5	55.4	54.3
10	84.5	77.5	72.3	68.2	65.0	62.4	60.1	58.3	56.5	55.1	53.7	52.5	51.5
11	82.8	75.7	70.3	66.3	62.9	60.1	58.0	56.0	54.4	52.7	51.3	50.1	49.2
12	81.5	74.0	68.6	64.4	61.2	58.3	55.8	54.0	52.4	50.9	49.3	48.2	47.2
13	80.0	72.5	67.2	62.9	59.4	56.7	54.4	52.2	50.5	48.9	47.7	46.6	45.4
14	79.0	71.5	65.7	61.5	58.1	55.1	52.8	50.8	49.0	47.5	46.2	45.0	44.0
15	78.1	70.3	64.6	60.3	57.0	53.9	51.5	49.4	47.8	46.1	44.9	43.6	42.5
16	77.3	69.4	63.5	59.2	55.6	52.9	50.3	48.3	46.6	45.2	43.6	42.4	41.4
17	76.5	68.4	62.6	58.2	54.7	51.8	49.3	47.1	45.5	44.0	42.6	41.4	40.3
18	75.5	67.6	61.8	57.1	53.7	50.8	48.4	46.3	44.6	43.0	41.6	40.4	39.3
19	75.1	67.0	60.9	56.5	52.8	49.9	47.4	45.5	43.7	42.1	40.7	39.6	38.4
20	74.1	66.1	60.1	55.6	52.1	49.1	46.6	44.7	42.9	41.3	39.8	38.7	37.5
25	72.0	63.3	57.1	52.5	48.9	46.0	43.4	41.4	39.6	38.1	36.7	35.5	34.5
30	69.8	61.1	55.1	50.3	46.7	43.6	41.1	39.1	37.3	35.7	34.4	33.2	32.1
35	68.5	59.6	53.3	48.6	44.9	41.9	39.5	37.4	35.6	34.0	32.7	31.6	30.4
40	67.5	58.3	52.0	47.3	43.4	40.5	38.0	36.0	34.1	32.7	31.3	30.1	29.1
45	66.4	57.1	50.8	46.1	42.4	39.3	36.9	34.8	33.1	31.5	30.2	29.0	27.9
50	65.6	56.3	49.9	45.2	41.4	38.4	35.9	33.9	32.1	30.5	29.2	28.1	27.0
100	60.9	51.4	44.9	40.0	36.3	33.3	31.0	28.9	27.2	25.8	24.5	23.3	22.3

TABLE A.4

95th percentile of the percentage of inertia explained by the first component of 10,000 PCAs performed on tables made up of independent variables (the number of individuals varies between 5 and 100, and the number of variables from 17 to 200). For example, for a table with $I = 50$ individuals and $K = 30$ variables, 95% of the percentages of inertia explained by the first plane are less than 19.1%.

						Number of variables							
nbind	17	18	19	20	25	30	35	40	50	75	100	150	200
5	74.9	74.2	73.5	72.8	70.7	68.8	67.4	66.4	64.7	62.0	60.5	58.5	57.4
6	67.0	66.3	65.6	64.9	62.3	60.4	58.9	57.6	55.8	52.9	51.0	49.0	47.8
7	61.3	60.7	59.7	59.1	56.4	54.3	52.6	51.4	49.5	46.4	44.6	42.4	41.2
8	57.0	56.2	55.4	54.5	51.8	49.7	47.8	46.7	44.6	41.6	39.8	37.6	36.4
9	53.6	52.5	51.8	51.2	48.1	45.9	44.4	42.9	41.0	38.0	36.1	34.0	32.7
10	50.6	49.8	49.0	48.3	45.2	42.9	41.4	40.1	38.0	35.0	33.2	31.0	29.8
11	48.1	47.2	46.5	45.8	42.8	40.6	39.0	37.7	35.6	32.6	30.8	28.7	27.5
12	46.2	45.2	44.4	43.8	40.7	38.5	36.9	35.5	33.5	30.5	28.8	26.7	25.5
13	44.4	43.4	42.8	41.9	39.0	36.8	35.1	33.9	31.8	28.8	27.1	25.0	23.9
14	42.9	42.0	41.3	40.4	37.4	35.2	33.6	32.3	30.4	27.4	25.7	23.6	22.4
15	41.6	40.7	39.8	39.1	36.2	34.0	32.4	31.1	29.0	26.0	24.3	22.4	21.2
16	40.4	39.5	38.7	37.9	35.0	32.8	31.1	29.8	27.9	24.9	23.2	21.2	20.1
17	39.4	38.5	37.6	36.9	33.8	31.7	30.1	28.8	26.8	23.9	22.2	20.3	19.2
18	38.3	37.4	36.7	35.8	32.9	30.7	29.1	27.8	25.9	22.9	21.3	19.4	18.3
19	37.4	36.5	35.8	34.9	32.0	29.9	28.3	27.0	25.1	22.2	20.5	18.6	17.5
20	36.7	35.8	34.9	34.2	31.3	29.1	27.5	26.2	24.3	21.4	19.8	18.0	16.9
25	33.5	32.5	31.8	31.1	28.1	26.0	24.5	23.3	21.4	18.6	17.0	15.2	14.2
30	31.2	30.3	29.5	28.8	26.0	23.9	22.3	21.1	19.3	16.6	15.1	13.4	12.5
35	29.5	28.6	27.9	27.1	24.3	22.2	20.7	19.6	17.8	15.2	13.7	12.1	11.1
40	28.1	27.3	26.5	25.8	23.0	21.0	19.5	18.4	16.6	14.1	12.7	11.1	10.2
45	27.0	26.1	25.4	24.7	21.9	20.0	18.5	17.4	15.7	13.2	11.8	10.3	9.4
50	26.1	25.3	24.6	23.8	21.1	19.1	17.7	16.6	14.9	12.5	11.1	9.6	8.7
100	21.5	20.7	19.9	19.3	16.7	14.9	13.6	12.5	11.0	8.9	7.7	6.4	5.7

A.2   R Software

A.2.1   Introduction

The R software is free and can be downloaded at the following address: `http://cran.r-project.org/`. The aim here is not to explain all of the different functions of the software, but rather to briefly outline how to conduct the analyses detailed in this work. For a more detailed presentation of R, please refer to the R manual:

`http://cran.r-project.org/doc/manuals/R-intro.html`

We will first describe a detailed example before moving on to list some of the most useful functions for importing data, constructing graphs, and so forth. In Section A.2.2, we present the Rcmdr package, which is used to conduct these analyses from a scroll-down menu, and in Section A.2.3, we present the FactoMineR package in further detail. This package is dedicated to data analysis and is used throughout this work. To begin, let us refer back to the example of the PCA on temperature data (see Section 1.10) and comment on the following lines of code:

```
1 > library(FactoMineR)
2 > temperature <- read.table("http://factominer.free.fr/bookV2/temperature.csv",
      header=TRUE,sep=";",dec=".",row.names=1)
3 > res <- PCA(temperature,ind.sup=24:35,quanti.sup=13:16,quali.sup=17)
4 > plot.PCA(res,choix="ind",habillage=17,cex=0.7,title="My PCA")
5 > plot.PCA(res, choix="var", select=c("June","Annual"))
6 > summary.PCA(res, nb.dec=2, nbelements=Inf)
7 > write.infile(res,file="c:/myfile.csv",sep=";")
```

1. Loading FactoMineR.

2. Importation from the dataset: the data table can be found in the file `http://factominer.free.fr/bookV2/temperature.csv`. The first line of the file contains the names of the variables; `sep=";"` the field separator is the character ";" (standard import format for csv files), `dec="."` the decimal separator is "."; `row.names=1` the first column contains the names of the individuals.

3. Conducting a PCA using the function **PCA**: individuals 24 to 35 (`24:35`) are supplementary, variables 13 to 16 are quantitative supplementary, and variable 17 is categorical supplementary. By default, the function centres and reduces the variables (the argument `scale.unit=TRUE` is used by default and does not need to be specified).

4. The function **plot.PCA** is vital to improve the default graphs: here, we colour-code the individuals according to the categories of variable 17 (categorical supplementary variable), character size is also

reduced (`cex=0.7` rather than 1 by default), and a title is given to the graph.

5. Construction of a graph of variables: the function **plot.PCA** enables us to choose the variables that we would like to appear on the graph of variables. Here, only the headings for the variables *June* and *Annual* are present.

6. The function **summary.PCA** returns the tables for the main results for the eigenvalues, the active individuals, the supplementary individuals, the active variables, and the quantitative or qualitative supplementary variables. Results are given with 2 decimal places for all the elements (`nbelement=Inf`, for *infinity*).

7. Exportation of the results: the function **write.infile** is used to write all of the results contained in the object `res` in a file (here in the file `c:/myfile.csv`).

Exporting graphs. The graphs can be exported in different formats (pdf, emf, eps, jpg, etc.). To choose the format, click on the graph and select `File` then `Save as`. Another option is to right-click on the graph and select `Copy as vectorial`. The graph can then be pasted directly into a word processing programme (Word or PowerPoint, for example). It is therefore possible to extract the graph and to modify it in order to improve its legibility (in PowerPoint, using Draw and Ungroup).

Choosing the individuals and/or variables for analysis. It is easy to conduct an analysis from part of a dataset. The following lines of code can be used to conduct a PCA on part of a data table (between the `[ , ]` the individuals are specified before the comma and the variables after the comma):

```
1 > res<-PCA(temperature[,1:12])
2 > res<-PCA(temperature[c(1:10,15:20),1:12])
3 > res<-PCA(temperature[-c(4:6,8,10),1:12])
```

1. On all the individuals but only with variables 1 to 12.

2. On individuals 1 to 10 and 15 to 20 but only with variables 1 to 12.

3. On all the individuals except 4, 5, 6, 8, and 10 and with variables 1 to 12.

Import and Export Functions

Function	Description
read.table	Imports a data table from a file and creates a data frame (table containing quantitative and/or categorical variables as well as information such as the names of the rows and columns)
read.csv	Imports a data table from a .csv file and creates a data frame

Function	Description
write.table	Writes a table into a file
write.infile	Function of the FactoMineR package which writes all of the elements of a list in a .csv file
save	Saves R objects in a `.Rdata` file
load	Finds the objects saved using the `save` function
history	Finds the most recently executed lines of code
save.history	Saves the history of the most recently executed lines of code

Data Management Functions

Function	Description
cbind.data.frame	Juxtaposes data frames into columns (groups the columns next to one another)
rbind.data.frame	Juxtaposes data frames into rows; the names of the columns from the data frames must be identical (groups the rows one on top of another, the columns are sorted in the same order for all of the tables in order to link the variables prior to merging)
sort	Sorts a vector in ascending order (or descending if `decreasing = TRUE`)
order	Sorts a table into one or more columns (or rows): `x[order(x[,3], -x[,6]), ]` sorts the table x in the (ascending) order of the third column in x then, in the case of equal values in the third column, in (decreasing) order of the sixth column of x
dimnames	Gives the names of the dimensions of an object (list, matrix, data frame, etc.)
rownames	Gives the names of the rows for a data frame or a matrix
colnames	Gives the names of the columns for a data frame or a matrix
dim	Gives the dimensions of an object
nrow	Gives the number of rows in a table
ncol	Gives the number of columns in a table
factor	Defines a vector as a factor, that is, a categorical variable (if `ordered=TRUE` the levels of the factors are considered to be in sequence)
levels	Gives the categories for a categorical variable (level of a factor)
nlevels	Gives the number of categories for a categorical variable

Function	Description
which	Gives the positions for the actual values of a vector or a logic table: the setting `arr.ind=TRUE` is used to return the numbers of the rows and columns in a table: **which**(`c(1,4,3,2,5,3)==3`) returns 3 and 6; **which**(**matrix**(`1:12,nrow=4)==3,arr.ind=TRUE`) returns (row 3, column 1)
is.na	Tests to see if the data is significant

Basic Statistical Functions

The following statistical functions are used to describe a quantitative variable x. For all of these functions, the setting `na.rm=TRUE` is used to eliminate the missing data prior to calculation. If `na.rm=FALSE` and the data are missing, the function will put out an error message.

Function	Description
mean(x, na.rm=TRUE)	Average of x
sd(x)	Standard deviation of x
var(x)	Variance of x if x is a variance–covariance vector or matrix if x is a matrix (unbiased variance)
cor(x)	Correlation matrix of x
quantile(x, probs)	Quantiles of x of the type `probs`
sum(x)	Sum of the elements of x
min(x)	Minimum of x
max(x)	Maximum of x
scale(x, center=TRUE, scale=TRUE)	Centres (`center=TRUE`) and reduces (`scale=TRUE`) x
colMeans(x)	Calculates the mean for each column in table x
rowMeans(x)	Calculates the mean for each row in table x
apply(x,MARGIN, FUN)	Applies the function FUN to the rows or columns in table x: **apply**(x, 2, mean) calculates the mean for each column in x; **apply**(x, 1, sum) calculates the sum for each row in x

Principal Component Functions

Function	Description
PCA	Principal component analysis with the possibility of including supplementary individuals, as well as quantitative and categorical supplementary variables

Function	Description
CA	Correspondence analysis with the possibility of including supplementary rows and columns
MCA	Multiple correspondence analysis with the possibility of including supplementary individuals, and quantitative and categorical supplementary variables
dimdesc	Describes the principal components (i.e., the dimensions)
catdes	Describes a categorical variable in terms of quantitative and/or categorical variables
condes	Describes a quantitative variable in terms of quantitative and/or categorical variables
HCPC	Hierarchical clustering on principal components
graph.var	Constructs the graph of variables from a limited number of variables

Graphic Functions

Function	Description
x11()	Creates a new, empty graph window
pdf, postscript, jpeg, png, bmp	Saves a graph in pdf, postscript, jpeg, png, or bmp format; all of the functions are used in the same way: **pdf("MyGraph.pdf")**; *graphic orders*; **dev.off()**

The Functions print, plot, and summary

The functions **print**, **plot**, and **summary** are generic functions; they give results that are unique depending on the class of the object to which they are applied.

Function	Description
print	Writes the results (all or an extract)
plot	Constructs a graph
summary	Gives the main results

For example, **print.PCA**, **print.CA**, **print.MCA**, can all be applied using the generic instruction **print**. Depending on the class of the object (output resulting from PCA, CA, MCA), the outputs or graphs will be specific to these analyses. To access assistance for the function of writing a given object, use, for example, PCA: help("print.PCA").

A.2.2 The Rcmdr Package

The graphic interface R Commander is available in the package Rcmdr. This interface means that R can be used simply, via a scroll-down menu. The aim of this package is also to help people to learn to use the software as it also provides the lines of code for the corresponding analyses. The Rcmdr interface does not contain all of the functions available in R, nor does it contain all of the options for the different functions, but the most common functions are programmed, and are thus available.

As with any package, it only needs to be installed once, and then loaded when needed, using

```
> library(Rcmdr)
```

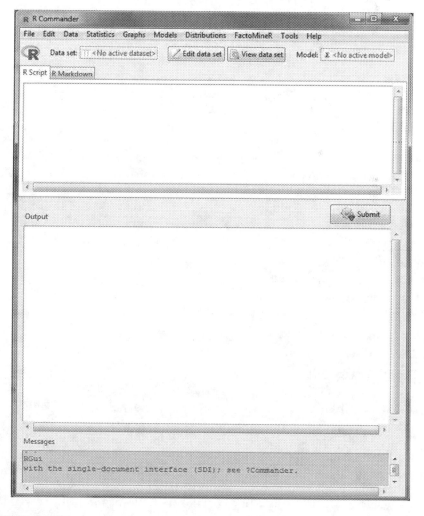

FIGURE A.1
Main window of Rcmdr.

The interface (see Figure A.1) opens automatically. This interface has a scroll-down menu, a script window, and an output window. When the scroll-down menu is used, the analysis is launched and the lines of code used to generate the analysis appear in the script window.

To import data with Rcmdr, the simplest option is to import a .txt or .csv file:

Data → Import data → from text file, clipboard or URL ...

The column separator (field separator) must then be specified, as must the decimal separator (a "." or a ",").

To verify that the dataset has been successfully imported:

Statistics → Summaries → Active data set

When importing a dataset in csv format which contains the individuals' logins, it is not possible to specify in Rcmdr's scroll-down menu that the first column contains the login. The dataset can be imported considering the login as a variable. The line of code is therefore modified in the the script window by adding the argument row.names=1 and then clicking on Submit.

To change the active dataset, click on the Data set box. If the active dataset is modified (for example, by converting a variable), this modification must be validated (=refresh) by

Data → Active data set → Refresh active data set

The output window writes the lines of code in red and the results in blue. The graphs are constructed in R. At the end of a Rcmdr session, the script window can be saved, including all of the instructions as well as the output window and thus the results. Both R and Rmcdr can be closed simultaneously by going to File → Exit → From Commander and R.

Remark

Writing in the Rcmdr script window or the R window amounts to the same thing. If an instruction is set in motion in Rcmdr, it will also be recognised in R and vice versa. Objects created by Rcmdr can therefore be used in R.

A.2.3 The FactoMineR Package

The FactoMineR package (Husson et al., 2009; Lê et al., 2008) is dedicated to exploratory data analysis. Most of the principal components methods are programmed within it: principal component analysis (**PCA** function), correspondence analysis (**CA** function), multiple correspondence analysis (**MCA** function), and hierarchical clustering on principal components (**HCPC** function). More advanced methods are also available and can be used to take into account structures relating to the variables or individuals. These additional methods are multiple factor analysis (**MFA** function), hierarchical multiple factor analysis (**HMFA** function), or dual multiple factor analysis (**DMFA** function). The function **catdes** is used to define a categorical variable according to quantitative and/or categorical variables. The function **condes** is used to define a quantitative variable according to quantitative and/or categorical variables. A brief description of these methods can be found in Lê et al. (2008).

For each method, one can also add supplementary elements: supplementary individuals, and supplementary quantitative and/or categorical variables. Many elements for facilitating interpretation are provided for each of these analyses: quality of representation, and contribution for individuals and variables. The graphical representations are at the heart of each of the analyses and there are a variety of available graphs: colour-coding the individuals according to a categorical variable, only representing the variables which are most successfully projected on the principal component map, and so on.

As with any package in R, it only needs to be installed once, and then loaded when needed, using

```
> library(FactoMineR)
```

There is a Web site entirely dedicated to the FactoMineR package: `http://factominer.free.fr`. It features all of the usage methods along with detailed examples.

Remark

Many packages for data analysis are available in R, in particular, the ade4 package. There is a Web site dedicated to this package, which provides a great number of detailed and commented examples: `http://pbil.univ-lyon1.fr/ADE-4`. There is another package on R which is entirely dedicated to clustering, be it hierarchical or otherwise, which is called cluster. It conducts the algorithms detailed in the work by Kaufman and Rousseuw (1990).[1]

The Scroll-Down Menu

A graphic interface is also available and can be installed as a user interface in the interface for the Rcmdr package (see Section A.2.2). To do so, the package RcmdrPlugin.FactoMineR must be installed once. Then, every time one wishes to use the FactoMineR scroll-down menu, Rcmdr must be loaded. To do so, click on Tools → Load Rcmdr plug-in(s) .... Choose the FactoMineR plug-in from the list; Rcmdr must then be restarted to take the new plug-in into account. If you want to have the FactoMineR scroll-down menu available in Rcmdr all the time, you can do Tools → Tools ==> Save Rcmdr options. The use of the scroll-down menu for PCA is detailed below.

1. Importing Data

    The Rcmdr scroll-down menu offers a number of formats for importing data. When the file is in text format (.txt, .csv), it is impossible to specify that the first column contain the individuals' identities (which is often the case in data analysis). It is therefore preferable to import using the FactoMineR menu

---

[1]Kaufman L. and Rousseuw P.J. (1990). *Finding Groups in Data. An Introduction to Cluster Analysis*, Wiley, New York, 342 p.

FactoMineR → Import data from txt file

Click on Row names in the first column (if the names of the individuals are present in the first column), and then specify the column separator (field separator) and the decimal separator.

2. PCA with FactoMineR

Click on the FactoMineR tab. Then, select Principal Component Analysis in order to open the main window of the PCA (see Figure A.2). It is possible to select supplementary categorical vari-

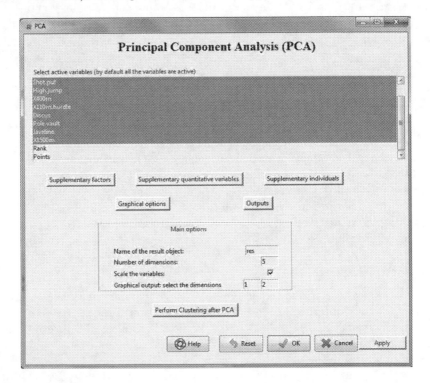

FIGURE A.2
Main window of PCA in the FactoMineR menu.

ables (Select supplementary factors), supplementary quantitative variables (Select supplementary variables), and supplementary individuals (Select supplementary individuals). By default, the results for the first five dimensions are provided in the object res, the variables are centred and reduced, and the graphics are provided for the first plane (components 1 and 2). It is preferable to choose Apply rather than Submit, since this means the window remains open whilst the analysis is being performed.

Certain options can therefore be modified without having to enter all of the settings a second time.

The window for graphical options (see Figure A.3) is separated into two parts. The left-hand part corresponds to the graph of individuals whereas the right-hand side refers to the graph of variables. It is possible to represent the supplementary categorical variables alone (without the individuals, under `Hide some elements:` select `ind`); it is also possible to omit the labels for the individuals (`Label for the active individuals`). The individuals can be colour-coded according to a categorical variable (`Colouring for individuals:` choose the categorical variable).

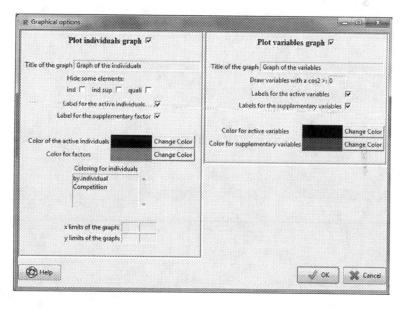

FIGURE A.3
Graphic options window in PCA.

The window for the different output options is used to choose how the different results are visualised (eigenvalues, individuals, variables, automatic description of the components). All of the results can also be exported in a `.csv` file (which can be opened using Excel).

Remark
The Factoshiny package also offers a graphical interface which enables users to construct interactive graphs directly from FactoMineR output. This interface can be used on the one hand to set parameters for the method, by choosing supplementary quantitative and qualitative variables in PCA, for example,

and on the other hand to improve graphs by altering font size, elements to be labelled, the identification of individuals according to the categories of a categorical variable, etc. See Section 5.5.1 for details.

Bibliography of Software Packages

The following is a bibliography of the main packages that perform exploratory data analysis or clustering in R. For a more complete list of packages, you can refer to the following link for exploratory data analysis methods:

http://cran.r-project.org/web/views/Multivariate.html

and the following one for clustering:

http://cran.r-project.org/web/views/Cluster.html

- *The ade4 Package* proposes data analysis functions to analyse Ecological and Environmental data in the framework of Euclidean Exploratory methods, hence the name ade4. The number of functions is very high and many functions can be used in framework other than the ecological one (functions **dudi.pca**, **dudi.acm**, **dudi.fca**, **dudi.mix**, **dudi.pco**, etc.).

  Dray S. & Dufour A.B. (2007). The ade4 package: Implementing the duality diagram for ecologists. *Journal of Statistical Software*, 22, 1–20.

  A Web site is dedicated to the package: http://pbil.univ-lyon1.fr/ADE-4

- *The ca Package*, proposed by Greenacre and Nenadic, deals with simple (function **ca**), multiple, and joint correspondence analysis (function **mjca**). Many new extensions for categorical variables are available in this package.

  Greenacre M. & Nenadic O. (2007). *ca: Simple, Multiple and Joint Correspondence Analysis*. R Package Version 0.21.

- *The cluster Package* allows us to perform basic clustering and particularly hierarchical clustering with the function **agnes**.

  Maechler M., Rousseeuw P., Struyf A., & Hubert M. (2005). *Cluster Analysis Basics and Extensions*.

- *The FactoMineR Package* is the one used in this book. It allows the user to perform exploratory multivariate data analyses (functions **PCA**, **CA**, **MCA**, **HCPC**) easily and provides many graphs (functions **plot**, **plotellipses**) and helps to interpret the results (functions **dimdesc**, **catdes**).

  Husson F., Josse J., Lê S., & Mazet J. (2009). *FactoMineR: Multivariate Exploratory Data Analysis and Data Mining with R*. R Package Version 1.14.

  Lê S., Josse J., & Husson F. (2008). FactoMineR: An R Package for multivariate analysis. *Journal of Statistical Software*, 25, 1–18.

A Web site is dedicated to the package: `http://factominer.free.fr`

- *The Factoshiny Package* is a visualisation software that has been initially developed for the FactoMineR package. The main objective of Factoshiny is to allow the user to explore graphical outputs interactively provided by multidimensional methods by visually integrating numerical indicators.

- *The homals Package* deals with homogeneity analysis. This is an alternative method to MCA for categorical variables. This method is often used in the psychometric community.

  De Leeuw J. & Mair P. (2009). Gifi methods for optimal scaling in R: The package homals. *Journal of Statistical Software*, 31(4), 1–20.

- *The hopach Package* builds (function **hopach**) a hierarchical tree of clusters by recursively partitioning a dataset, while ordering and possibly collapsing clusters at each level.

- *The MASS Package* allows the user to perform some very basic analyses. The functions **corresp** and **mca** perform correspondence analysis.

  Venables W.N. & Ripley B.D. (2002). *Modern Applied Statistics with S*, 4th ed., Springer, New York.

- *The missMDA Package* allows the user to perform imputation for missing values with multivariate data analysis methods, for example, according to a PCA model or MCA model. Combined with the FactoMineR package, it allows users to handle missing values in PCA and MCA.

- *The R Software* has some functions to perform exploratory data analysis: **princomp** or **prcomp**, **hclust**, **kmeans**, **biplot**. These functions are very basic and offer no help for interpreting the data.

  R Foundation for Statistical Computing. (2009). *R: A Language and Environment for Statistical Computing*, Vienna, Austria.

- *The Rcmdr Package* proposes a graphical user interface (GUI) for R. Many basic methods are available and moreover, several extensions are proposed for specific methods, for example, RcmdrPlugin.FactoMineR.

- Murtagh F. (2005). proposes correspondence analysis and hierarchical clustering code in R. `http://www.correspondances.info`

Bibliography

This bibliography is divided in several sections to partition the references according to the different methods: Principal Component Analysis, Correspondence Analysis, and Clustering Methods.

References on All the Exploratory Data Methods

- Escofier B. & Pagès J. (2008). *Analyses Factorielles Simples et Multiples: Objectifs, Méthodes et Interprétation*, 4th ed. Dunod, Paris.

- Gifi A. (1981). *Non-Linear Multivariate Analysis*. D.S.W.O.-Press, Leiden.

- Govaert G. (2009). *Data Analysis*. Wiley, New York.

- Lê S., Josse J., & Husson F. (2008). FactoMineR: An R package for multivariate analysis. *Journal of Statistical Software*, 25(1), 1–18.

- Le Roux B. & Rouanet H. (2004). *Geometric Data Analysis, from Correspondence Analysis to Structured Data Analysis*. Kluwer, Dordrecht.

- Lebart L., Morineau A., & Warwick K. (1984). *Multivariate Descriptive Statistical Analysis*. Wiley, New York.

- Lebart L., Piron M., & Morineau A. (2006). *Statistique Exploratoire Multidimensionnelle: Visualisation et Inférence en Fouilles de Données*, 4th ed. Dunod, Paris.

References for Chapter 1: Principal Component Analysis

- Gower J.C. & Hand D.J. (1996). *Biplots*. Chapman & Hall/CRC Press, London.

- Jolliffe I.T. (2002). *Principal Component Analysis*, 2nd ed. Springer, New York.

References for Chapter 2: Correspondence Analysis and for Chapter 3: Multiple Correspondence Analysis

- Benzécri J.P. (1973). *L'analyse des Données*, Tome 2 Correspondances. Dunod, Paris.

- Benzécri J.P. (1992). *Correspondence Analysis Handbook* (Transl.: T.K. Gopalan). Marcel Dekker, New York.

- Greenacre M. (1984). *Theory and Applications of Correspondence Analysis*. Academic Press, London.

- Greenacre M. (2007). *Correspondence Analysis in Practice*. Chapman & Hall/CRC Press, London.

- Greenacre M. & Blasius J. (2006). *Multiple Correspondence Analysis and Related Methods*. Chapman & Hall/CRC Press, London.

- Le Roux B. & Rouannet H. (2010). *Multiple Correspondence Analysis*. Series: Quantitative Applications in the Social Sciences. Sage, London.

- Lebart L., Salem A., & Berry L. (2008). *Exploring Textual Data*. Kluwer, Boston.

- Murtagh F. (2005). *Correspondence Analysis and Data Coding with R and Java*. Chapman & Hall/CRC Press, London.

References for Chapter 4: Clustering Methods

- Hartigan J. (1975). *Clustering Algorithms*. Wiley, New York.

- Kaufman L. & Rousseuw P. (1990). *Finding Groups in Data. An Introduction to Cluster Analysis*. Wiley & Sons, New York.

- Lerman I.C. (1981). *Classification Automatique et Ordinale des Données*. Dunod, Paris.

- Mirkin B. (2005). *Clustering for Data Mining: A Data Recovery Approach*. Chapman & Hall/CRC Press, London.

- Murtagh F. (1985). *Multidimensional Clustering Algorithms*. COMPSTAT Lectures, Physica-Verlag, Vienna.

Index

Printed in the United States
by Baker & Taylor Publisher Services